THE NORTHWEST EUROPEAN POLLEN FLORA, I

THE NORTHWEST EUROPEAN POLLEN FLORA, I

Parts 1—7

Editor

W. Punt   Utrecht (The Netherlands)

Coeditors

C.R. Janssen   Utrecht (The Netherlands)
Tj. Reitsma   Lelystad (The Netherlands)
G.C.S. Clarke London (Great Britain)

Published under the auspices of the Royal Botanical Society of The Netherlands

Reprinted from *Review of Palaeobotany and Palynology*,
Volumes 17, 19 and 21

ELSEVIER SCIENTIFIC PUBLISHING COMPANY
Amsterdam — Oxford — New York   1976

ELSEVIER SCIENTIFIC PUBLISHING COMPANY
335 Jan van Galenstraat
P.O. Box 211, Amsterdam, The Netherlands

Distributors for the United States and Canada:

ELSEVIER NORTH-HOLLAND INC.
52, Vanderbilt Avenue
New York, N.Y. 10017

ISBN: 0-444-41421-5

Copyright © 1976 by Elsevier Scientific Publishing Company, Amsterdam

All rights reserved. No part of this publication may be reproduced, stored in a retrieval system, or transmitted in any form or by any means, electronic, mechanical, photocopying, recording, or otherwise, without the prior written permission of the publisher,
Elsevier Scientific Publishing Company, Jan van Galenstraat 335, Amsterdam

Printed in The Netherlands

CONTENTS

1. Introduction
   C.R. Janssen, W. Punt and Tj. Reitsma (Utrecht, The Netherlands) . . 1
2. Caprifoliaceae
   W. Punt, Tj. Reitsma and A.A.M.L. Reuvers (Utrecht,
   The Netherlands) 5
3. Primulaceae
   W. Punt, J.S. de Leeuw van Weenen and W.A.P. van Oostrum
   (Utrecht, The Netherlands) 31
4. Adoxaceae
   Tj. Reitsma and A.A.M.L. Reuvers (Utrecht, The Netherlands) . . 71
5. Sparganiaceae and Typhaceae
   W. Punt (Utrecht, The Netherlands) 75
6. Gentianaceae
   W. Punt and W. Nienhuis (Utrecht, The Netherlands) 89
7. Guttiferae
   G.C.S. Clarke (London, Great Britain) 125
   Index 143
   Errata 145

The Northwest European Pollen Flora

INTRODUCTION

C. R. JANSSEN, W. PUNT and Tj. REITSMA*

Laboratory of Palaeobotany and Palynology, State University, Utrecht (The Netherlands)

INTRODUCTION

In modern pollen analysis it has become clear that palaeoecological and phytogeographical interpretations are much enhanced by a detailed determination of pollen types. Frequently it proves impossible to key out the fossil types to the species level, but identification of the pollen grains to taxa of the lowest possible rank is highly desirable. Quaternary palynologists in Western Europe have a number of excellent pollen keys at their disposal, such as: (1) key to the NW European pollen types by Faegri and Iversen (1950, 1964); (2) Leitfaden der Pollenbestimmung, 1 by Beug (1961); and (3) pollen and spore key by Nilsson and Praglowski (1963). These keys, however, are not always sufficiently detailed. On the other hand, keys treating special taxonomic or other groups are available, such as: (1) Chanda (1963) — Caryophyllaceae; (2) Praglowski (1963) — Swedish trees; and (3) Cerceau-Larrival (1959) — Umbelliferae; etc. The number of these publications, however, is restricted and without continuity. In order to fill the gap we are planning to start publishing a "Northwest European Pollen Flora" on a regular basis.

We intend to study pollen grains of all seed plants and the spores of a number of important pteridophytes and bryophytes. The main goal of this project is to publish keys to the pollen types and to illustrate them by means of photographs for easy reference. Moreover, this Northwest European Pollen Flora will also include comprehensive descriptions of the pollen types and information on the pertinent literature. Much attention will be paid to variability in pollen populations, and — if possible — to occurrence in Quaternary deposits. The taxonomic position, evolution, and palaeogeography of the pollen types and taxa will be stressed to a lesser extent.

Information on the taxa will come from monographs and the most recent floras available from the specific area. Nomenclature follows the Flora Europaea except where stated to the contrary.

MATERIAL

In principle all the material to be used for the preparation of the pollen slides originates from a specific area in Europe shown on the map (Fig.1).

---

*Present address: Rijksdienst voor de IJsselmeerpolders, Lelystad (The Netherlands).

Roughly it covers most of the area occupied by the Atlantic and general European plant-geographical elements. However, we shall not be too dogmatic in this respect.

Not all plants growing in Northwest Europe will be included in the pollen flora and the choice will be made arbitrarily. It was decided that garden plants and adventitious plants will be excluded from the flora. Introduced species are only listed if they have been established in a sizable part of the area. Neophytes, i.e. introduced plants which are completely naturalized, will certainly be studied as they often play an important role in the present vegetation (e.g. *Prunus serotina, Galinsoga parviflora*). Also some species not occurring in Northwest Europe today, but sometimes found in Quaternary deposits, such as *Myrtus communis, Tsuga, Brasenia*, etc., will be taken into consideration.

If a species is poorly represented in Northwest Europe, material from outside the area will be examined and also species occurring at the borders of the area and not distinctly confined to the Alpine region or the Mediterranean area are to be considered.

The source of the plant material is primarily the herbarium of the State University of Utrecht and, besides, the Rijksherbarium at Leiden. In order to avoid errors because of misidentifications of the herbarium specimens every specimen will be re-identified and cited.

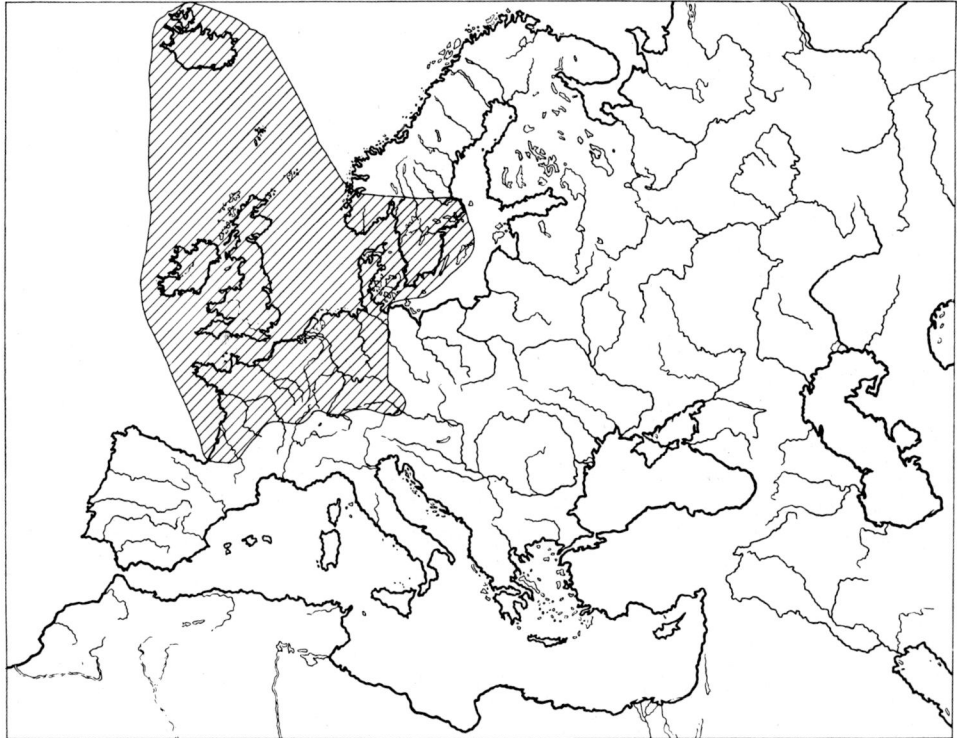

Fig.1. Map of the area dealt with in the Northwest European Pollen Flora (hatched).

Information regarding herbarium specimens in the publication will be restricted to specific name, author, collector's name and number, herbarium, and country. Other data are available in the files of the Laboratory of Palaeobotany and Palynology at Utrecht. At least two collections from different localities within the area will be examined, but usually more specimens will be studied in order to trace possible specific and intraspecific variability.

METHODS

Preparation method

Pollen grains will be prepared according to the acetolysis method as described by Reitsma (1969). The mounting is in: (a) glycerine jelly (Reitsma, 1969), as well as in (b) silicone oil (Andersen, 1960), because both mounting methods are in use with palynologists.

Terminology

For descriptions the terminology as suggested by Reitsma (1970) is used. If necessary, however, terms not mentioned or explained by Reitsma will be defined separately in the publications.

Descriptions

The pollen grains will be examined with Leitz microscopes and Leitz objectives. The sequence of pollen characteristics in the descriptions is always given in the same order. The following sequence of groups of features is used: *pollen class*, *apertures*, *exine*, *ornamentation*, *outlines*, and *measurements*. After a description a secondary key to pollen groups or species may follow and sometimes a brief comment can be added.

Measurements will be made both in preparations embedded in silicone oil and those mounted in glycerine jelly. They are often of relative value for a detailed identification of fossil pollen types (Reitsma, 1969) and only by a strict statistical treatment measurements may be of use for this purpose. We do not intend, however, to treat measurements statistically; they will be presented in a rather simple way by extremes only. They are considered to be of relative value.

Keys

The keys will differentiate between the species as much as possible with the use of the optics mentioned above, either on the level of pollen types or of pollen groups. In the publications a pollen type is defined as the lowest pollen-morphological entity, morphologically separated from other types by one or more distinct features. If the differences are less clear, pollen groups are involved.

A main key will lead to the pollen types. The features in this key may usually be observed with the aid of a 40X objective. Secondary keys may identify the smaller pollen groups or species, often with the aid of objectives of stronger magnification and resolution.

Photographs

Photographs illustrating the pollen types will be taken with a Leitz Ortholux microscope combined with an Orthomat camera. It is our intention to picture at least all pollen types with their characteristics and variation and, if possible, also the pollen groups and different species. Usually a number of SEM-microphotographs will be added.

Publications

The Flora is a publication under the auspices of the Royal Botanical Society of the Netherlands. Each section deals with a single family, and the series has its own pagination. These sections will not appear in a strict systematic order or in a regular sequence. Though the family treatments are in many cases of special interest to Quaternary palynologists, the authors express the hope that the Flora will prove its value to those interested in pollen morphology and plant taxonomy.

REFERENCES

Andersen, S. T., 1960. Silicone oil as mounting medium for pollen grains. Dan. Geol. Unders., IV (1): 1—24.
Beug, H. J., 1961. Leitfaden der Pollenbestimmung, 1. Fischer, Stuttgart, 92 pp.
Cerceau-Larrival, M. T., 1959. Clé de détermination d'Ombellifères de France et d'Afrique du Nord d'après leurs grains de pollen. Pollen Spores, 1: 145—190.
Chanda, S., 1963. On the pollen morphology of some Scandinavian Caryophyllaceae. Grana Palynol., 3: 67—89.
Faegri, K. and Iversen, J., 1950. Textbook of Modern Pollen Analysis. Munksgaard, Copenhagen, 169 pp.
Faegri, K. and Iversen, J., 1964. Textbook of Pollen Analysis. Munksgaard, Copenhagen, 2nd ed., 237 pp.
Nilsson, S. and Praglowski, J. R., 1963. Pollen and spore key. In: G. Erdtman, J. Praglowski and S. Nilsson (Editors), An Introduction to a Scandinavian Pollen Flora, I. Almquist and Wiksell, Uppsala, 92 pp.
Praglowski, J. R., 1963. Notes on the pollen morphology of Swedish Trees and shrubs. In: G. Erdtman, J. R. Praglowski and S. Nilsson (Editors), An Introduction to a Scandinavian Pollen Flora, II. Almquist and Wiksell, Uppsala, 89 pp.
Reitsma, Tj., 1969. Size modifications of recent pollen grains under different treatments. Rev. Palaeobot. Palynol., 9: 175—202.
Reitsma, Tj., 1970. Suggestions towards unification of descriptive terminology of angiosperm pollen grains. Rev. Palaeobot. Palynol., 10: 39—60.

The Northwest European Pollen Flora, 2

CAPRIFOLIACEAE

W. PUNT, Tj. REITSMA and ALBERDINA A. M. L. REUVERS

Laboratory of Palaeobotany and Palynology, State University, Utrecht (The Netherlands)

LITERATURE

Bassett and Compton (1970), Erdtman (1943, 1952, 1966), Erdtman et al. (1961, 1963), Faegri and Iversen (1950, 1964), Kuprianova and Alyoshina (1972), Radulescu (1961), Richard (1970).

TERMINOLOGY

Bridge (Faegri and Iversen, 1950, 1964, emend.): the margins of the colpi are raised in the equatorial area and connected with each other; this connection forms a bridge over the colpus dividing the ectoaperture into two parts (Plate III, 1).

Fastigium (Reitsma, 1966, 1970, emend.): cavity in a colporate grain, caused by a separation of the inner part of the exine and the domed outer part of the exine in the area of the endoaperture.

Retipilate(Erdtman, 1952): pollen grain provided with a reticulum, whereby the muri consist of detached pila.

SPECIMENS EXAMINED

Linnaea borealis L. — Finland: Nordstrøm s.n.(U); The Netherlands: de Smidt 32(U); Norway: Biol. exc. 1965—2506(U), Brummer-De Vries s.n.(U).
Lonicera alpigena L. — Switzerland: Biol. exc. 1946—203(U), Biol. exc. 1965—1399(U), Mennega s.n.(U), Simon Thomas 1299(U).
Lonicera caprifolium L. — Germany: Behrendsen 1375(U); Yugoslavia: Biol. exc. 1961—302(U).
Lonicera coerulea L. — Italy: Biol. exc. 1956—145(U); Switzerland: Biol. exc. 1922 s.n.(U), Biol. exc. 1962—3057(U), Mennega s.n.(U).
Lonicera nigra L. — Austria: Behrendsen 1380(U), Kramer 1347(U); Italy: Biol. exc. 1955—156(U); Yugoslavia: Behrendsen s.n.(U).
Lonicera periclymenum L. — Great Britain: Biol. exc. 1972—2134(U); The Netherlands: Gadella and Mennega s.n.(U), Kramer s.n.(U), Reitsma s.n.(U); Sweden: Hekking 3485(U).
Lonicera xylosteum L. — Finland: Haakana s.n.(U); France: Kramer and Westra 3683(U); Sweden: Blom s.n.(U); Switzerland: Biol. exc. 1965—1351(U).
Sambucus ebulus L. — Germany: Biol. exc. 1964—2377(U); The Netherlands: Kok-Ankersmit s.n.(U), Mennega 671(U), Punt (fresh material); Portugal: Rainha 4311(U); Turkey: De Koster et al. s.n.(U).
Sambucus nigra L. — The Netherlands: Dieleman 48(U), Kramer and Westra s.n.(U), Punt (fresh material), Rem(fresh material), s.c. s.n.(HbU—046321) (U); Sweden: Hekking 3415(U).
Sambucus racemosa L. — Austria: Traxler 519(U); France: Biol. exc. 1965—1117(U); The Netherlands: Van der Burgh s.n.(U), Punt(fresh material), Van Royen 49(U); Siberia: Maximowicz s.n.(U).
Viburnum lantana L. — Czechoslovakia: Svestka 1368(U); France: Biol. exc. 1953—330(U).

Viburnum opulus L. — Finland: Haakana s.n.(U); The Netherlands: Leeuwenberg 254(U), Punt(fresh material), Rem(fresh material).
Viburnum tinus L. — Israel: Zohary 76(U); Portugal: Fernandes 4321(U).

KEY TO THE POLLEN TYPES

1.a. Pollen grains reticulate 2
  b. Pollen grains not reticulate 6
2.a. Reticulum coarse, lumina larger than 2μ, exine thick, up to 4μ 3
  b. Reticulum fine, lumina ca. 1μ, exine thin, up to 2.5μ ....
.... *Sambucus nigra* type (see also Adoxaceae)
3.a. Sexine 1 in mesocolpium thicker than in apocolpium ....
....*Sambucus ebulus* type
  b. Sexine 1 in mesocolpium as thick as in apocolpium 4
4.a. Costae endocolpi distinct, bridge absent .... *Viburnum tinus* type
  b. Costae endocolpi absent, bridge absent or, if present, narrow or broad
5.a. P/E ratio about or smaller than 1, ornamentation in apocolpium not reticulate, in mesocolpium reticulate or retipilate ....
.... *Viburnum lantana* type
  b. P/E ratio larger than 1, ornamentation in apo- and mesocolpium reticulate .... *Viburnum opulus* type
6.a. Pollen grains scabrate .... *Linnaea borealis* type
  b. Pollen grains echinate 7
7.a. Sexine 1 + sexine 2 much thicker than nexine, columellae in surface view distinct, longest axis usually larger than 60μ 8
  b. Sexine 1 + sexine 2 about as thick as nexine, columellae in surface view rather indistinct, longest axis usually smaller than 60μ 10
8.a. Echinae up to 2.5μ, ends of the ectocolpi usually obtuse ....
.... *Lonicera alpigena* type
  b. Echinae up to 1.5μ, ends of the ectocolpi usually acute 9
9.a. Costae of the endoaperture present all around the endoaperture, longest axis of the endoaperture ca. 36μ ....*Lonicera periclymenum* type
  b. Costae of the endoaperture only present at the polar sides of the endoaperture, longest axis of the endoaperture ca. 27μ ....
....*Lonicera caprifolium* type
10.a. Echinae dimorphic: large, widely spaced ones mixed with small ones; pollen grains usually 4-zonocolporate .... *Lonicera coerulea* type
  b. Echinae monomorphic: large, widely spaced; pollen grains usually 3-zonocolporate .... *Lonicera xylosteum* type

DESCRIPTION OF THE POLLEN TYPES

Linnaea borealis type (Plate XI, 2—8)

Pollen class: 3-(4-)Zonocolporate.
P/E ratio: Subtransverse to suberect.
Apertures: Ectoaperture — colpus, of medium length to short, narrow, sunken; margins of the colpus undulate; end obtuse; fastigium. Endoaperture —

lolongate porus; longest axis of the endoporus longer than the longest axis of the ectocolpus; distinct costae at the polar sides of the endoporus tapering towards the mesocolpium or absent towards the mesocolpium.
Exine: Thick; sexine always thicker than nexine; nexine thinner than sexine 1; sexine 1 thicker than sexine 2 (tectum); sexine 3, consisting of scabrae, thinner than sexine 2.
Ornamentation: Supra-scabrate; scabrae densely spaced; structure of sexine 1 rather indistinct.
Outlines: Equatorial view — circular to elliptic. Polar view — circular or 3- or 4-angular with convex sides and apertures situated in the obtuse angles.
Measurements: Glycerine jelly — P 40—50μ; E 40—50μ; P/E 0.94—1.09; apocolpium index 0.68—0.74; exine about 4μ. Silicone oil — P 39—47μ; E 39—47μ; P/E 0.85—1.
Species: Linnaea borealis

Comments
This type is characterized by the scabrae, which are never found in the other types of the Caprifoliaceae.

Lonicera alpigena type (Plate V, 1—8)

Pollen class: 3- or 4-Zonocolporate.
P/E ratio: Subtransverse to adequate.
Apertures: Ectoaperture — colpus, short and narrow, sunken; margins of the colpus undulate; end obtuse; mostly with fastigium. Endoaperture — lalongate porus; shortest axis of the endoporus as long as or slightly shorter than the longest axis of the ectocolpus; costae at the polar sides of the endoporus, tapering towards its ends, absent in the mesocolpium.
Exine: Thick; sexine always thicker than nexine; sexine 1 thicker than sexine 2 (tectum); nexine thinner than sexine 1 + sexine 2; sexine 3, consisting of firm echinae, thinner than nexine + sexine 1 + sexine 2; sometimes the inner face of the nexine undulate.
Ornamentation: Supra-echinate; echinae widely spaced, thickened at the base; structure of sexine 1 quite distinct.
Outlines: Equatorial view — circular to elliptic. Polar view — 3- or 4-angular with convex sides and apertures situated in the obtuse angles.
Measurements: Glycerine jelly — E (longest axis) 69—81μ; P 63—75μ; P/E 0.89—1; apocolpium index 0.62—0.89; exine about 5.5μ; echinae up to 2.5μ; longest axis of the endoporus about 20μ; shortest axis of the endoporus about 17μ. Silicone oil — E (longest axis) 63—77μ; P 54—66μ; P/E 0.78—0.95.
Species: Lonicera alpigena

Comments
This type is intermediate between the *Lonicera coerulea* and the *Lonicera xylosteum* type on the one side and the *Lonicera caprifolium* and *Lonicera periclymenum* type on the other side. It differs from the *Lonicera coerulea* and the *Lonicera xylosteum* type on the basis of the nexine which is thinner

than the sexine 1 + sexine 2 and from the *Lonicera caprifolium* and *Lonicera periclymenum* type on the basis of the large and firm echinae.

Lonicera caprifolium type (Plate VI, 1—9)

Pollen class: (2-), 3-Zonocolporate.
P/E ratio: Semitransverse to subtransverse.
Apertures: Ectoaperture — colpus, of medium length to short, narrow, sunken; margins of the colpus undulate; end acute; fastigium. Endoaperture — lalongate, truncate elliptic porus; shortest axis of the endoaperture about as long as the longest axis of the ectocolpus; thick costae at the polar sides of the endoaperture, in mesocolpium rather indistinct.
Exine: Thick; sexine always thicker than nexine; nexine thinner than sexine 1 + sexine 2; sexine 1 thicker than sexine 2(tectum); sexine 3, consisting of echinae, thinner than sexine 1 + sexine 2.
Ornamentation: Supra-echinate; echinae widely spaced; not thickened at the base; structure of sexine 1 distinct.
Outlines: Equatorial view — elliptic, obtusely acuminate. Polar view — 3-angular with convex sides and apertures situated in the obtuse angles.
Measurements: Glycerine jelly — E (longest axis) 68—77μ; P 60—72μ; P/E ratio 0.83—0.95; apocolpium index 0.71—0.80; exine about 5μ; echinae up to 1.5μ; longest axis of the endoaperture 27μ; shortest axis of the endoaperture 20μ. Silicone oil — E (longest axis) 66—77μ; P 63—72μ; P/E ratio 0.87—0.97.
Species: Lonicera caprifolium

Comments
This type has strong affinities to the *Lonicera periclymenum* type. The difference between these two types lies in the costae of the endoapertures. In this type the costae endopori are confined to the polar part of the endopori, whilst in the *Lonicera periclymenum* type the costae endopori are present all around the entire margin of the endoporus.

Lonicera coerulea type (Plate VII, 1—8)

Pollen class: (3-), 4-, (5-) Zonocolporate.
P/E ratio: Semitransverse to subtransverse.
Apertures: Ectoaperture — colpus, of medium length to short, narrow, sunken; margins of the colpus undulate; end obtuse; fastigium. Endoaperture — lalongate porus; shortest axis of the endoporus as long as or somewhat shorter than the longest axis of the ectocolpus; costae rather thin, all around the endoporus.
Exine: Thick; sexine always thicker than nexine; nexine thicker than sexine 1 + sexine 2; sexine 1 about as thick as sexine 2(tectum); sexine 3, consisting of echinae, as thick as or thicker than nexine + sexine 1 + sexine 2; inner face of the nexine sometimes undulate.
Ornamentation: Supra-echinate; echinae dimorphic, i.e., widely spaced

firm echinae, which are thickened at the base, and densely spaced small ones; structure of sexine 1 small and fine, densely spaced.
Outlines: Equatorial view — elliptic sometimes obtusely acuminate or truncate. Polar view — 3- or 4- or 5-angular with convex to straight sides and apertures situated in the obtuse angles.
Measurements: Glycerine jelly — E (longest axis) 45—57μ; P 34—55μ; P/E 0.76—0.96; apocolpium index 0.68—0.88; exine about 5μ; firm echinae up to 3μ; small echinae up to 1.5μ; longest axis of the endoaperture about 14μ; shortest axis of the endoaperture about 11μ. Silicone oil — E (longest axis) 42—57μ; P 39—57μ; P/E 0.82—1.0.
Species: Lonicera coerulea

Comments
This type has affinities to the *Lonicera xylosteum* type, but differs in the shape and the length of the echinae. Echinae in this type are dimorphic, i.e., widely spaced firm and densely spaced small echinae occur, while the *Lonicera xylosteum* type has firm echinae only.

Lonicera periclymenum type (Plate X, 1—8; Plate XI, 1; Plate XII, 3)

Pollen class: 3-(4-)Zonocolporate.
P/E ratio: Semitransverse to subtransverse.
Apertures: Ectoaperture — colpus, of medium length to short, narrow, sunken; margins of the colpus undulate; end acute; fastigium. Endoaperture — lalongate porus; thick costae, somewhat tapering towards the ends; shortest axis of the endoporus somewhat shorter than the longest axis of the ectocolpus.
Exine: Thick; sexine always thicker than nexine; nexine thinner than sexine 1 + sexine 2; sexine 1 thicker than sexine 2 (tectum); sexine 3, consisting of echinae, thinner than sexine 1 + sexine 2.
Ornamentation: Supra-echinate; echinae widely spaced, not thickened at the base; structure of sexine 1 distinct.
Outlines: Equatorial view — elliptic, obtusely acuminate. Polar view — 3- or 4-angular with convex to straight or, rarely, concave sides and apertures situated in the obtuse angles.
Measurements: Glycerine jelly — E (longest axis) 69—83μ; P 63—76μ; P/E ratio 0.85—0.97; apocolpium index 0.70—0.80; exine about 5μ; echinae up to 1.5μ; longest axis of the endoporus about 36μ; shortest axis of the endoporus about 19μ. Silicone oil — E 69—81μ; P 67—75μ; P/E ratio 0.86—0.98.
Species: Lonicera periclymenum

Comments
This type has strong affinities to the *Lonicera caprifolium* type.
For the differentiating characters see the comments on the *Lonicera caprifolium* type (p.8).

Lonicera xylosteum type (Plate IX, 1—8)

Pollen class: (2-), 3-, (4-)Zonocolporate.
P/E ratio: Semitransverse to suberect.
Apertures: Ectoaperture — colpus, of medium length to short, narrow, sunken; margins of the colpus undulate; end obtuse; fastigium. Endoaperture — either lalongate porus or, rarely, lalongate colpus; costae rather thin, present only in the polar part of the endoaperture; ends of the endoaperture in the mesocolpium rather indistinct; shortest axis of the endoaperture as long as or slightly shorter than the longest axis of the ectocolpus.
Exine: Thick; sexine always thicker than nexine; nexine thinner than or as thick as sexine 1 + sexine 2; sexine 1 as thick as or thicker than sexine 2 (tectum); sexine 3, consisting of firm echinae, as thick as or thicker than nexine + sexine 1 + sexine 2; sometimes the inner face of the nexine undulate.
Ornamentation: Supra-echinate; widely spaced firm echinae thickened at the base; structure of sexine 1 distinct or indistinct.
Outlines: Equatorial view — circular to elliptic. Polar view — circular to 3- or 4-angular with convex to straight sides and then apertures situated in the obtuse angles.
Measurements: Glycerine jelly — E (mostly the longest axis) 46—66μ; P (sometimes the longest axis) 45—60μ; P/E ratio 0.80—1.05; apocolpium index 0.57—0.80; exine about 5μ; echinae up to 3μ; longest axis of the endoaperture 16—23μ; shortest axis of the endoaperture 10—21μ. Silicone oil — E 49—60μ; P 42—57μ; P/E ratio 0.82—1.05.
Species: Lonicera nigra, Lonicera xylosteum

Key to the species
1.a. Distance between columellae very short, always smaller than 0.5μ; columellae hardly visible in surface view; shortest axis of the endoaperture ca. 11μ ....*L. nigra*
   b. Distance between columellae short, ca. 0.5μ; columellae visible in surface view; shortest axis of the endoaperture about 17μ..*L. xylosteum*

L. nigra
(2-), 3-Zonocolporate; endoporus or endocolpus, longest axis of the endoaperture about 19μ, shortest axis about 11μ; sexine 1 structure rather indistinct; undulate nexine at the inner face of the exine; outline in polar view circular to 3-angular with slightly convex or straight sides.

L. xylosteum
3-, (4-)Zonocolporate; endoporus, longest axis of the endoaperture about 21μ, shortest axis about 17μ; structure of sexine 1 distinct; outline in polar view circular or 3- or 4-angular with convex sides.

Comments
This type shows affinities to the *Lonicera coerulea* type. For the

differences between these two types see the comments on the *L. coerulea* type (p.*9*).

Sambucus ebulus type (Plate III, 2—6; Plate IV, 1—2; Plate XII, 4, 5)

Pollen class: 3-Zonocolporate.
P/E ratio: Subtransverse to semi-erect.
Apertures: Ectoaperture — colpus, long and wide, deeply sunken; end acute; fastigium; bridge indistinct; membrane nudate. Endoaperture — lalongate colpus; indistinct costae.
Exine: Thick; sexine distinctly thicker than nexine; sexine 1 distinctly thicker than sexine 2 in mesocolpium, sexine 1 decreasing towards the apocolpium, sexine 1 as thick as sexine 2 in apocolpium; capita small, in optical section slightly broader than columellae, circular.
Ornamentation: Reticulate; lumina of the reticulum in mesocolpium large, irregularly shaped, mostly 5- to 6-angular; lumina in apocolpium small, within the lumina small bacula or granules; muri simplicolumellate, narrow; columellae in surface view circular or slightly elliptic.
Outlines: Equatorial view — elliptic. Polar view — 3-angular with convex sides and apertures situated in the obtuse angles.
Measurements: Glycerine jelly — P $21-28\mu$; E $21-27\mu$; P/E ratio 0.90—1.30; apocolpium index 0.18—0.28; exine about 3μ; lumina in mesocolpium up to 3μ, in apocolpium smaller than 1μ. Silicone oil — P $23-26\mu$; E $21-25\mu$; P/E ratio 1.00—1.13.
Species: Sambucus ebulus

Comments
This type shows some affinities to the *Viburnum lantana* type, the *V. opulus* type, and the *V. tinus* type. The differential characters of these types are listed in Table I.

Sambucus nigra type (Plate IV, 3—10)

Pollen class: (2-), 3-Zonocolporate.
P/E ratio: Suberect to erect.
Apertures: Ectoaperture — colpus, long and wide, deeply sunken; end acute to slightly obtuse; bridge either distinct or indistinct and narrow; membrane nudate. Endoaperture — lalongate colpus, rather indistinct without costae.
Exine: Thin; sexine as thick as or thicker than nexine; decreasing in thickness towards the colpi; sexine 1 varying in thickness.
Ornamentation: Reticulate; lumina irregularly shaped, rather small in mesocolpium; lumina decreasing in size towards the colpi and the apocolpium; muri simplicolumellate; columellae in surface view circular.
Outlines: Equatorial view — circular to elliptic. Polar view — circular or triangular with convex sides and apertures situated in the obtuse angles.
Measurements: Glycerine jelly — P (longest axis) $19-26\mu$; E $15-22\mu$; P/E

ratio 1.02—1.50; apocolpium index 0.12—0.26; exine about 1.5μ; lumina up to 1μ. Silicone oil — P (longest axis) 18—22μ; E 17—21μ; P/E ratio 1.02—1.19.
Species: Sambucus nigra, Sambucus racemosa

Key to the species
1.a. Inner face of the exine of the mesocolpium convex in polar view; sexine 1 as thick as or thinner than sexine 2; capita in optical section distinctly broader than the rather indistinct columellae; bridge narrow, indistinct .... *S. racemosa*
  b. Inner face of the exine of the mesocolpium straight in polar view; sexine 1 thicker than sexine 2; capita in optical section small and spherical, based on rather distinct columellae; bridge distinct .... *S. nigra*

Comments
This type differs from all other types in the genera *Sambucus* and *Viburnum* by the thin exine and the fine reticulum. However, it matches very nearly the *Adoxa moschatellina* type, from which it differs only in the nudate membrane of the ectocolpus and the lalongate endocolpus (Reitsma and Reuvers, 1974).

Viburnum lantana type (Plate II, 4—7; Plate III, 1)

Pollen class: (2-), 3-, (4-)Zonocolporate.
P/E ratio: Semitransverse to subtransverse.
Apertures: Ectoaperture — colpus, long and wide, deeply sunken; end obtuse; bridge broad, distinct; membrane nudate. Endoaperture — lalongate colpus, narrow, rather indistinct without costae.
Exine: Thick; sexine as thick as or slightly thinner than nexine; sexine 1 about as thick as sexine 2, usually decreasing in thickness towards the ectocolpi; capita in optical section distinctly broader than columellae.
Ornamentation: Reticulate to retipilate in mesocolpium, pilate in apocolpium; lumina in mesocolpium not completely closed, irregularly shaped, slightly decreasing towards the ectocolpi and set with small but distinct granules or bacula; muri simplicolumellate, laterally not always completely fused; size of columellae in surface view variable, angular, circular or elliptic.

TABLE I

Differential characters of the *Sambucus ebulus* type and the *Viburnum* types

	S. ebulus type	*V. lantana* type	*V. opulus* type	*V. tinus* type
ornamentation at apocolpium	reticulate	pilate	reticulate	reticulate
costae endocolpi	+	—	—	+
columellae decreasing towards the poles	+	—	—	—

Present +, absent —.

Outlines: Equatorial view — circular to elliptic. Polar view — 3-angular with convex sides and apertures situated in the obtuse angles.
Measurements: Glycerine jelly — E (longest axis) 21—30μ; P 18—28μ; P/E ratio 0.85—0.98; apocolpium index 0.15—0.22; exine about 3μ; lumina up to 3.5μ. Silicone oil — E (longest axis) 25—28μ; P 24—27μ; P/E ratio 0.84—0.98.
Species: Viburnum lantana

Comments
This type shows some affinities to the *Sambucus ebulus* type, the *Viburnum opulus* type and the *V. tinus* type. The differential characters of these types are shown in Table I.

Viburnum opulus type (Plate I, 1—6; Plate XII, 1—2)

Pollen class: (2-), 3-, (4-)Zonocolporate.
P/E ratio: Suberect to semi-erect.
Apertures: Ectoaperture — colpus, long and wide, deeply sunken; end slightly obtuse; bridge either broad or narrow, distinct; membrane nudate. Endoaperture — lalongate colpus, rather indistinct without costae.
Exine: Thick; sexine as thick as or slightly thicker than nexine; sexine 1 about as thick as sexine 2; capita distinctly broader than columellae, circular in optical section.
Ornamentation: Reticulate; lumina wide, irregularly shaped, more or less angular, decreasing towards the ectocolpi, set with small granules or bacula; muri simplicolumellate; columellae circular or slightly elliptic in surface view.
Outlines: Equatorial view — elliptic. Polar view — 3-angular with convex sides and apertures situated in the obtuse angles.
Measurements: Glycerine jelly — P (longest axis) 25—30μ; E 23—26μ; P/E ratio 1.05—1.30; apocolpium index about 0.2; exine up to 3.5μ; lumina up to 3.5μ. Silicone oil — P (longest axis) 24—29μ; E 21—24μ; P/E ratio 1.13—1.18.
Species: Viburnum opulus

Comments
This type shows affinities to the *Sambucus ebulus* type, the *Viburnum lantana* type, and the *V. tinus* type. The differential characters of these types are listed in Table I.

Viburnum tinus type (Plate II, 1—3)

Pollen class: (2-), 3-, (4-)Zonocolporate.
P/E ratio: Suberect to erect.
Apertures: Ectoaperture — colpus, long and wide, deeply sunken; end acute; margins of the colpus not completely connected into a bridge but nearly so; fastigium; membrane nudate. Endoaperture — lalongate colpus; costae.
Exine: Thick; sexine slightly thicker than nexine; sexine 1 distinctly thicker

than sexine 2; in optical section capita small, only slightly broader than the columellae, circular in shape.
Ornamentation: Reticulate; lumina wide, irregularly shaped, decreasing towards the ectocolpi, set with small granules or bacula; muri simplicolumellate; columellae circular to elliptic in surface view.
Outlines: Equatorial view — elliptic. Polar view — 3-angular with slightly convex sides and apertures situated in the obtuse angles.
Measurements: Glycerine jelly — P (longest axis) $26-39\mu$; E $21-30\mu$; P/E ratio $1.07-1.42$; apocolpium index $0.13-0.24$; exine about 3μ; lumina up to 4μ. Silicone oil — P (longest axis) $26-37\mu$; E $23-32\mu$; P/E ratio $1.18-1.25$.
Species: Viburnum tinus

Comments
This type shows affinities to the *Sambucus ebulus* type, the *Viburnum lantana* type and the *V. opulus* type. The differential characters are shown in Table I.

PLATE DESCRIPTIONS

PLATE I (× 1500; p.*17*)

Viburnum opulus L. (Haakana s.n.)
1. Outline in polar view.
2. Outline in equatorial view.
3. Apocolpium with outer ends of the colpi.
4. Reticulum at low focus; small granules within the lumina.
5. Outline in equatorial view with endoaperture.
6. Reticulum at high focus.

PLATE II (× 1500; p.*18*)

Viburnum tinus L. (Zohary 762)
1. Outline in polar view.
2. Endocolpus.
3. Apocolpium with outer ends of the colpi.
Viburnum lantana L. (Biol. exc. 1953—330)
4. Apocolpium.
5. Reticulum; columellae irregular in outline.
6. Outline in equatorial view.
7. Outline in polar view; thick nexine.

PLATE III (× 1500; p.*19*)

Viburnum lantana L. (Biol. exc. 1953—330)
1. Bridge, showing two parts of the colpus.
Sambucus ebulus L. (Kok-Ankersmit s.n.)
2. Ectocolpus.
3. Outline in polar view.
4. Outline in equatorial view of normally expanded pollen grain.
5. Colpus with endoaperture and indistinct costae.
6. Outline in equatorial view of aberrant pollen grain; columellae lower at the poles.

PLATE IV (× 1500; p.*20*)

Sambucus ebulus L. (Kok-Ankersmit s.n.)
1. Reticulum at high focus.
2. Reticulum at low focus, granules within the lumina.

Sambucus nigra L. (Dieleman 48)
3. Colpus with bridge.
4. Outline in polar view; inner face of nexine straight.
5. Outline in equatorial view of aberrant pollen grain.
6. Reticulum at high focus.
7. Reticulum at low focus.
8. Outline in equatorial view of normal pollen grain.

Sambucus racemosa L. (Traxler U 19)
9. Apocolpium.
10. Outline in polar view; inner face of nexine convex.

PLATE V (× 1500, except 6; p.*21*)

Lonicera alpigena L. (Thomas 1299)
1—5. Ornamentation from high focus to low focus.
6. Outline in polar view (× 1200).
7. Endoaperture.
8. Ectoaperture.

PLATE VI (× 1500, except 5; p.*22*)

Lonicera caprifolium L. (Biol. exc. 1960—302)
1—4. Ornamentation from high focus to low focus.
5. Outer end of endoaperture (× 800).
6. Ectoaperture.
7. Detail of outline in polar view (mesocolpium).
8. Ectoaperture.
9. Endoaperture.

PLATE VII (× 1500, except 1, 3; p.*23*)

Lonicera coerulea L. (Mennega s.n.)
1. Outline in equatorial view (× 800).
2. Endoaperture with thin costae.
3. Outline in polar view (× 800).
4. Detail of outline in polar view (mesocolpium).
5—8. Ornamentation from high focus to low focus.

PLATE VIII (× 1500, except 1; p.*24*)

Lonicera nigra L. (Behrendsen 1380)
1. Outline in polar view (× 800).
2—5. Ornamentation from high focus to low focus.
6. Ectoaperture.
7. Endoaperture with rather thin costae.
8. Detail of outline in polar view (mesocolpium).

PLATE IX (× 1500, except 1; p.*25*)

Lonicera xylosteum L. (Kramer and Westra 3683)
1. Outline in polar view (× 925).
2. Ectoaperture.
3. Endoaperture with rather thin costae.
4—8. Ornamentation from high focus to low focus.

PLATE X (× 1500, except 7; p.*26*)

Lonicera periclymenum L. (Hekking 3485)
1. Ectoaperture.
2. Outer ends of the endoaperture.
3—6. Ornamentation from low focus to high focus.
7. Outline in polar view (× 800).
8. Endoaperture with thick costae.

PLATE XI (× 1500, except 1, 5—8; p.*27*)

Lonicera periclymenum L. (Hekking 3485)
1. Outline in equatorial view (× 800).
Linnaea borealis L. (Nordström s.n.)
2. Outline in polar view.
3. Endoaperture with distinct costae at the polar sides.
4. Ectoaperture.
5—8. Ornamentation from high focus to low focus (× 1200).

PLATE XII (Scanning electron micrographs; p.*28*)

Viburnum opulus L. (Haakana s.n.)
1. Equatorial view; colpus with small, narrow bridge (× 2750).
2. Structure of mesocolpial reticulum; lumina with numerous granules (× 5000).
Lonicera periclymenum L. (Hekking 3485)
3. Equatorial view (× 1000).
Sambucus ebulus L. (Mennega 671)
4. Equatorial view; muri high; columellae long (× 3500).
5. Structure of mesocolpial reticulum; lumina with few granules (× 5000).

PLATE I

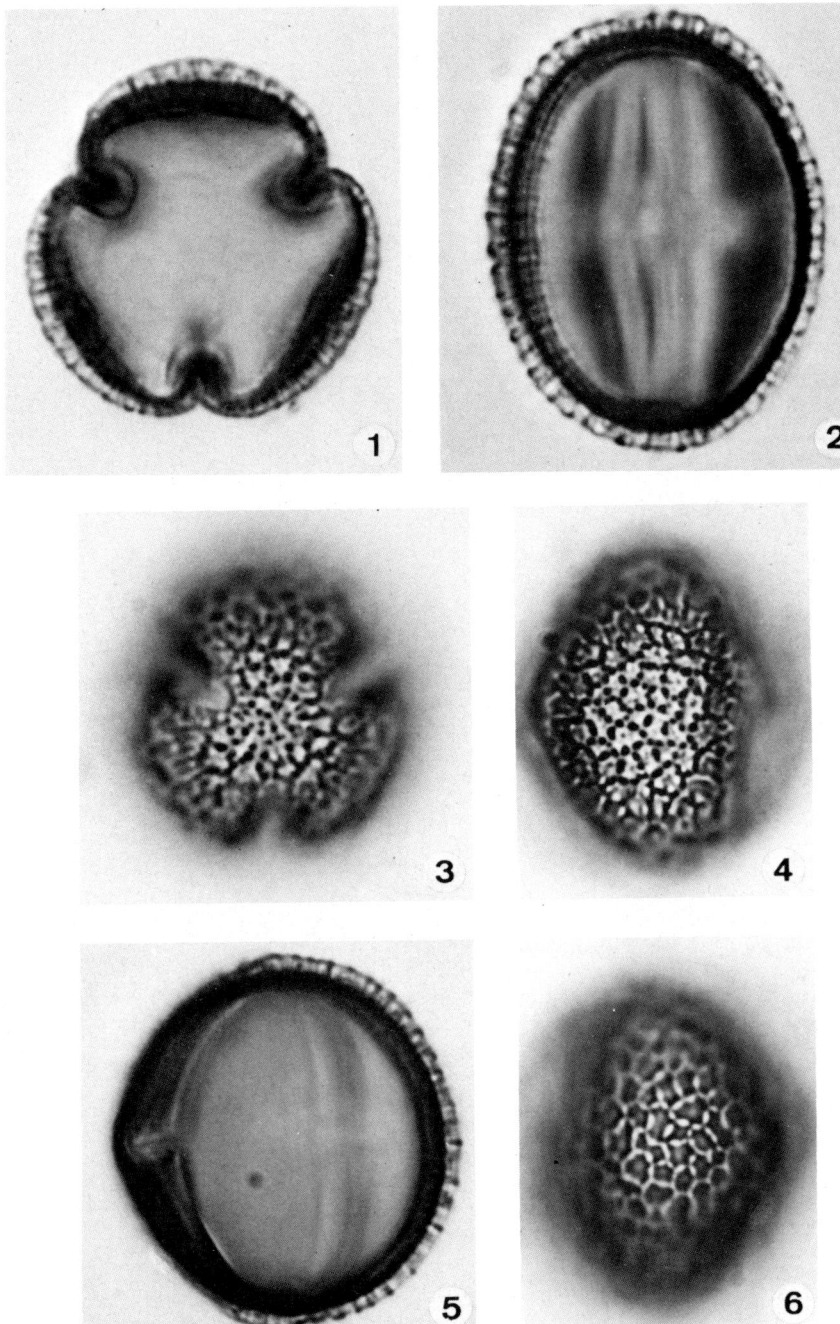

PLATE II

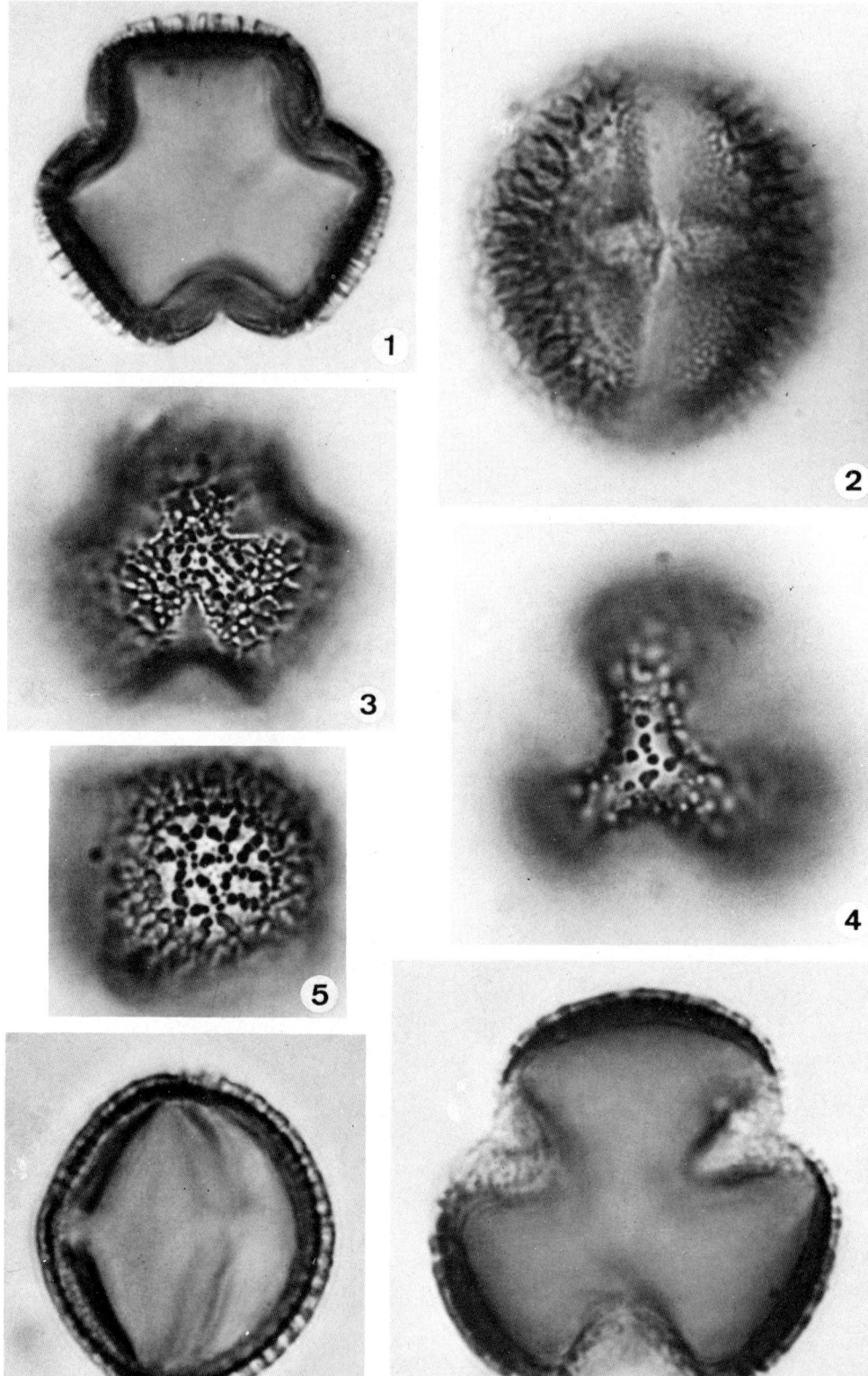

PLATE III

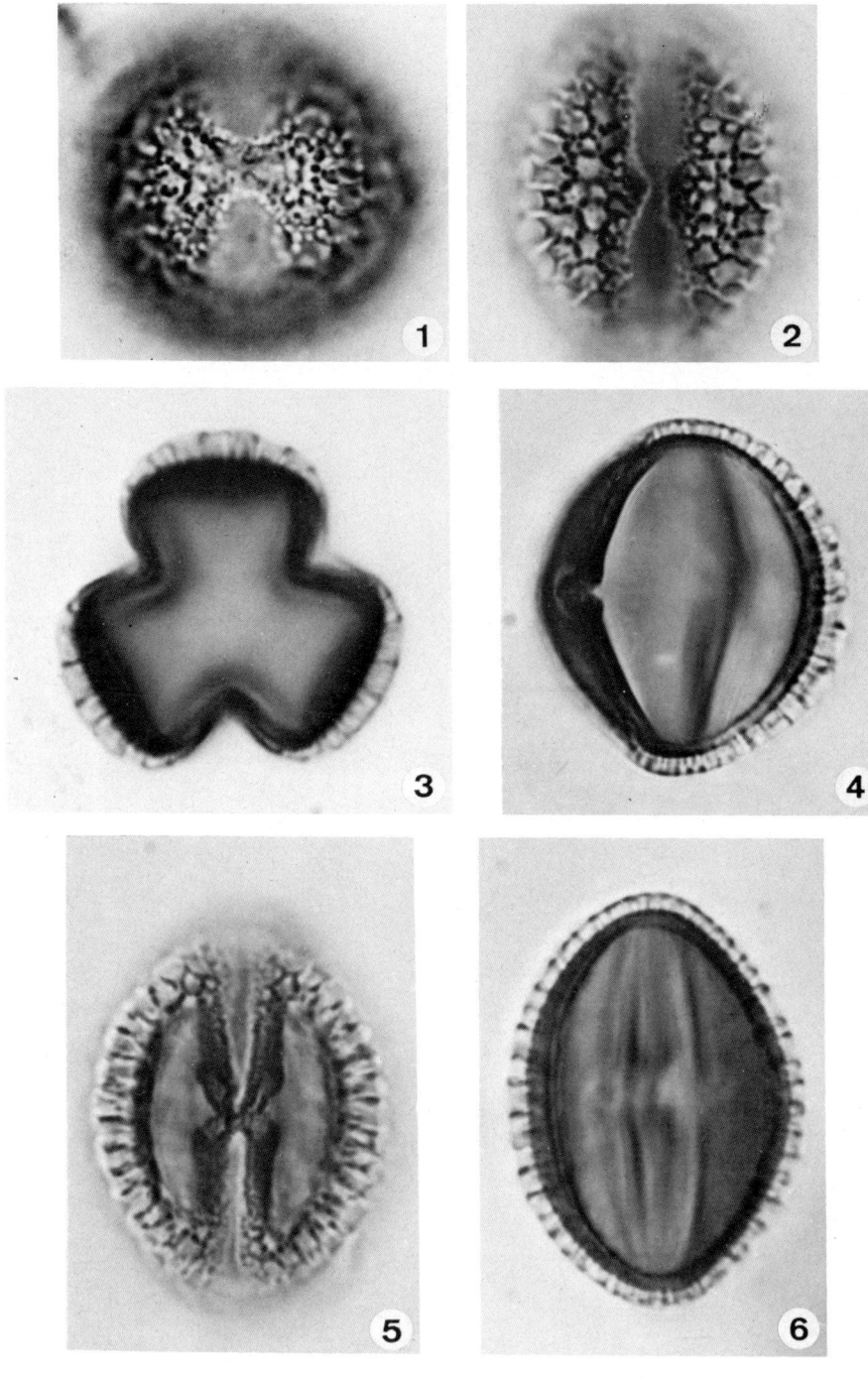

20

PLATE IV

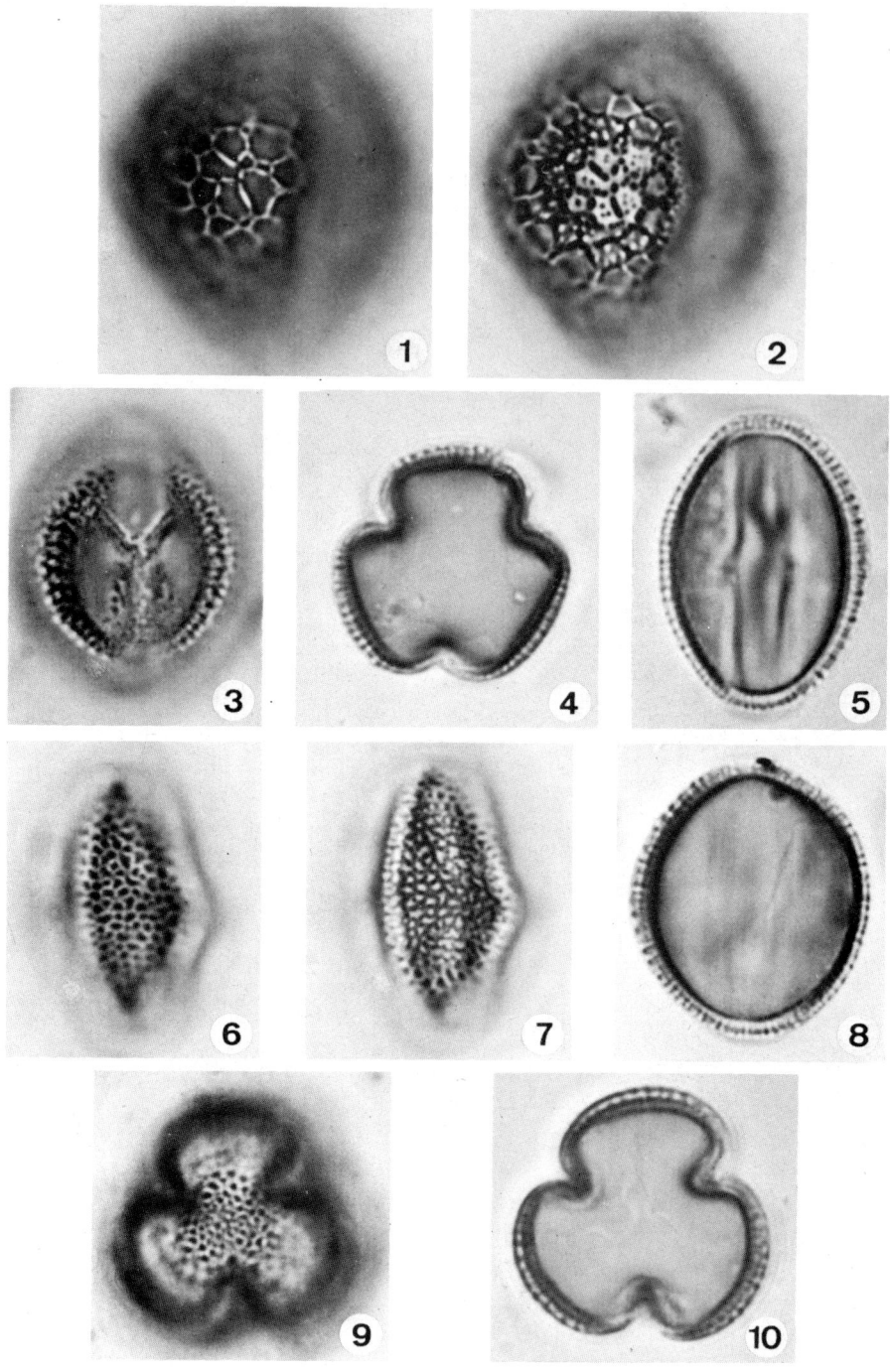

PLATE V

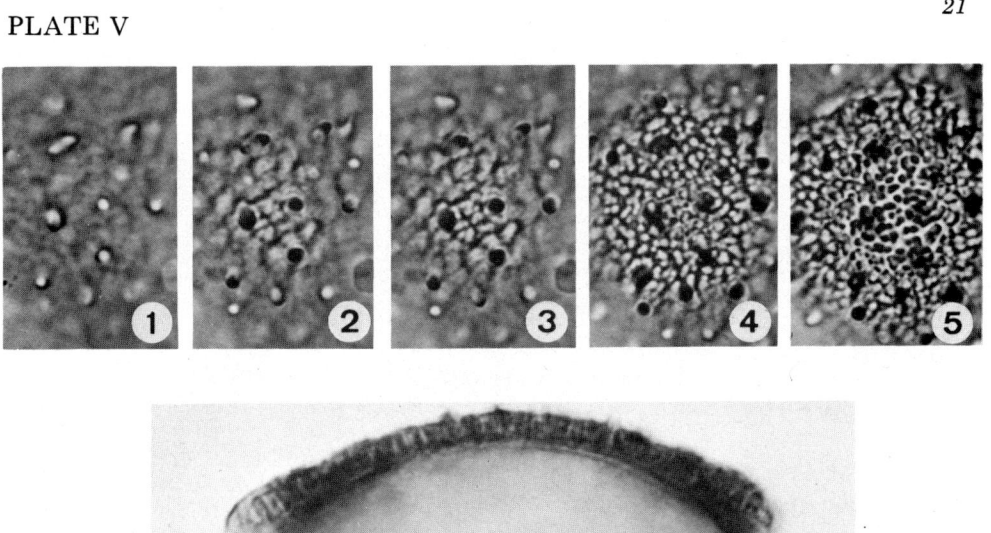
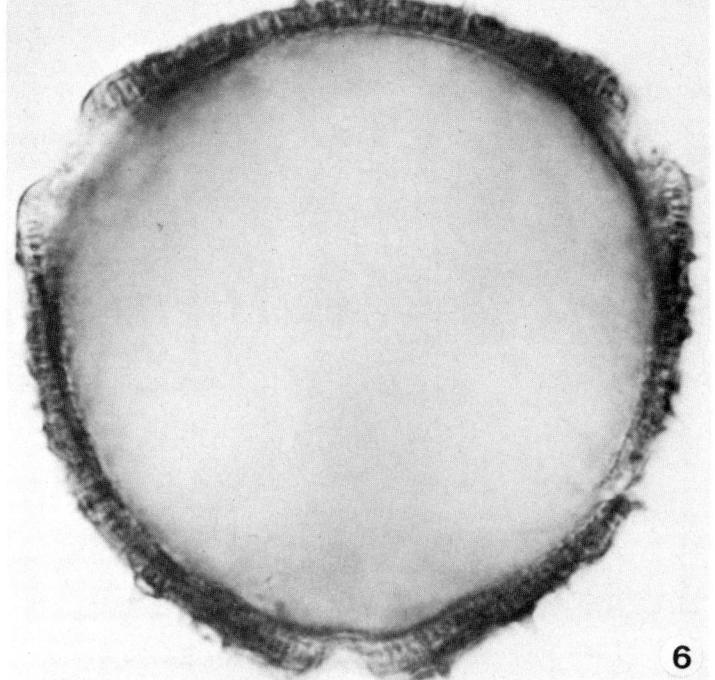
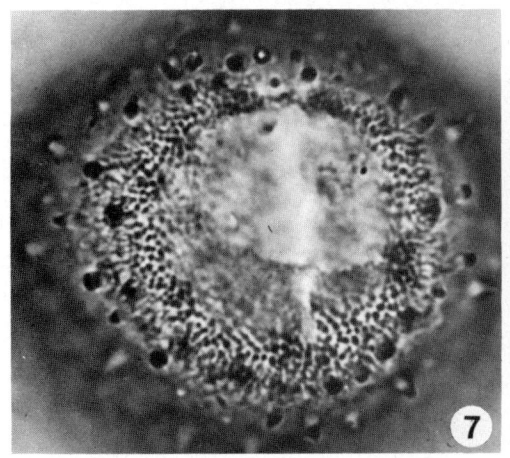
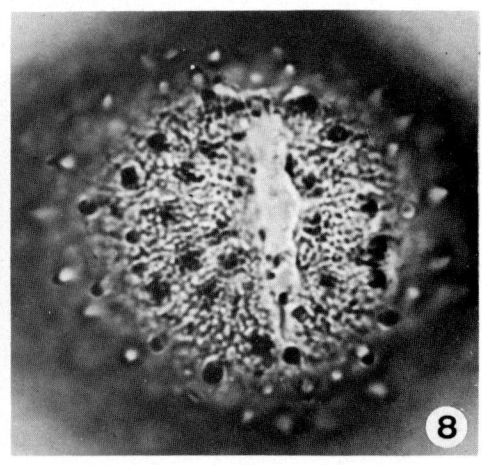

PLATE VI

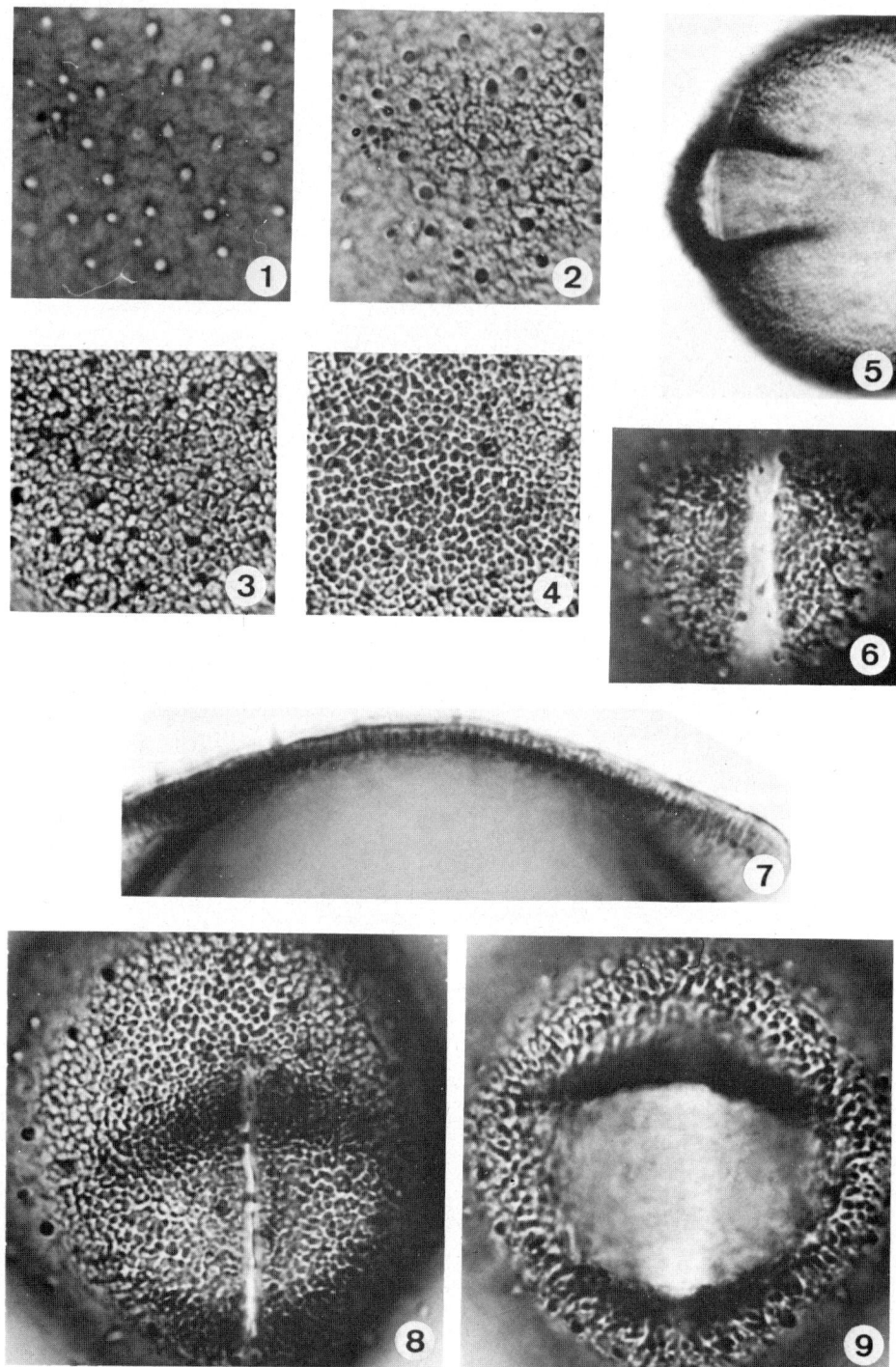

PLATE VII

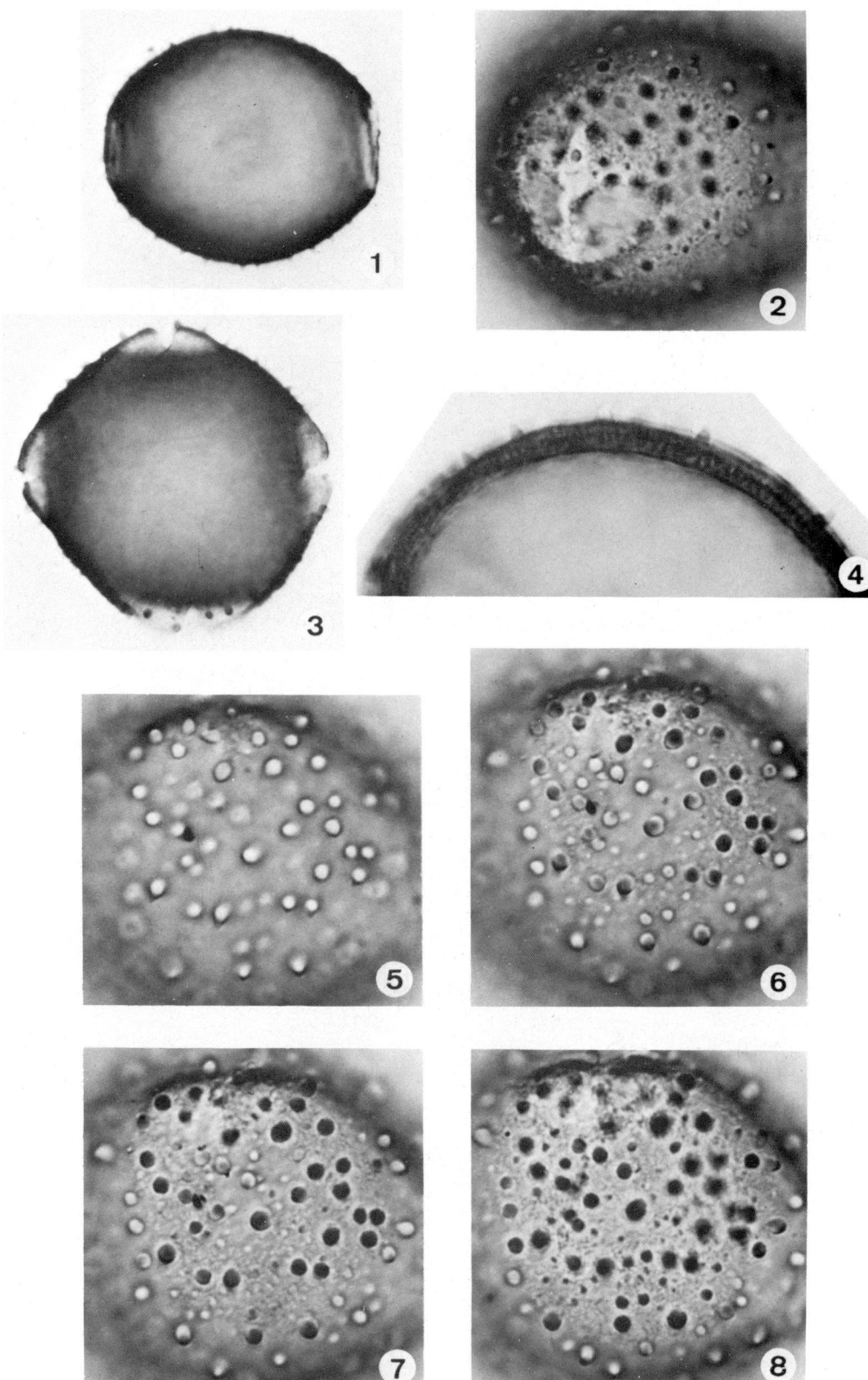

PLATE VIII

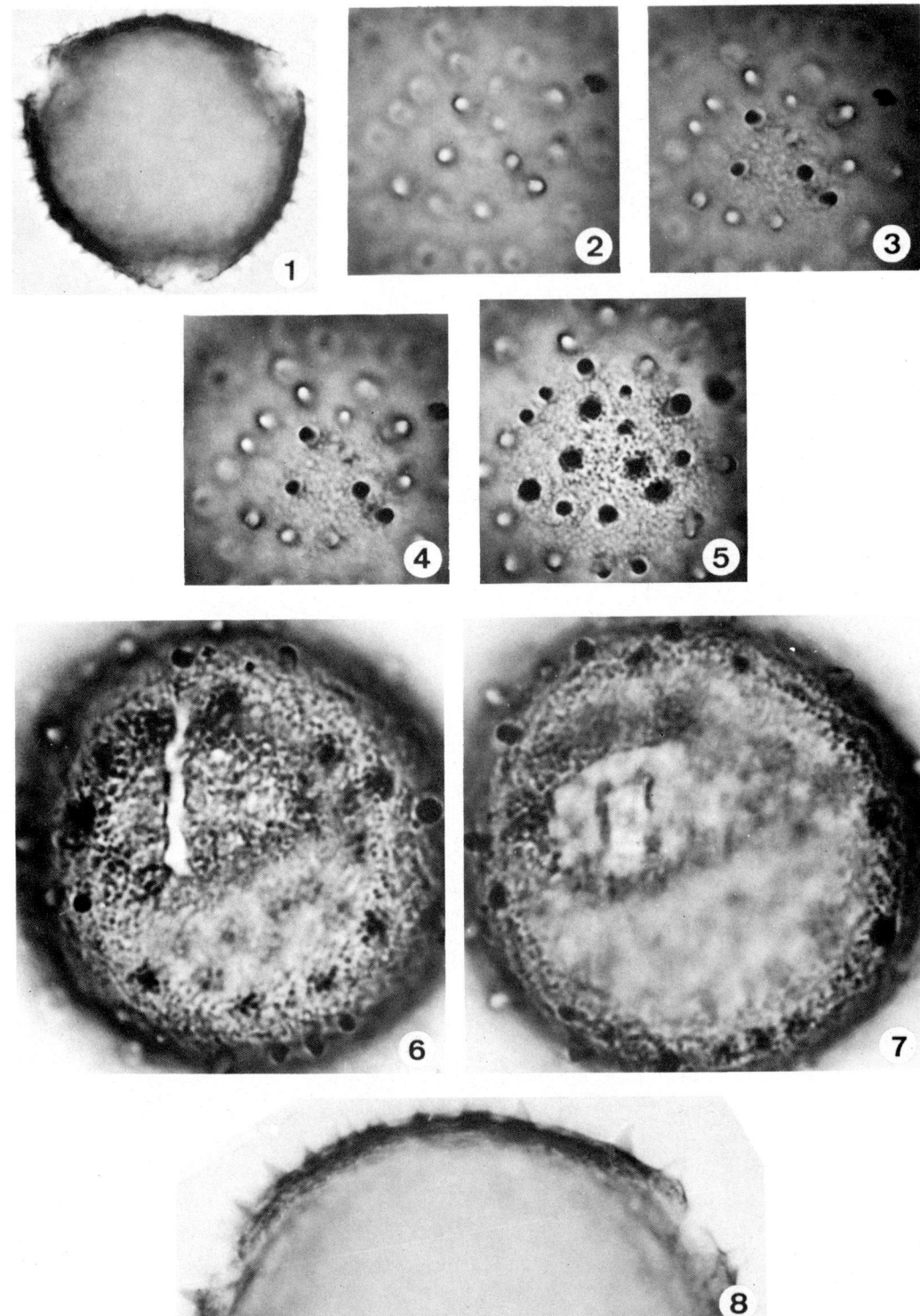

PLATE IX

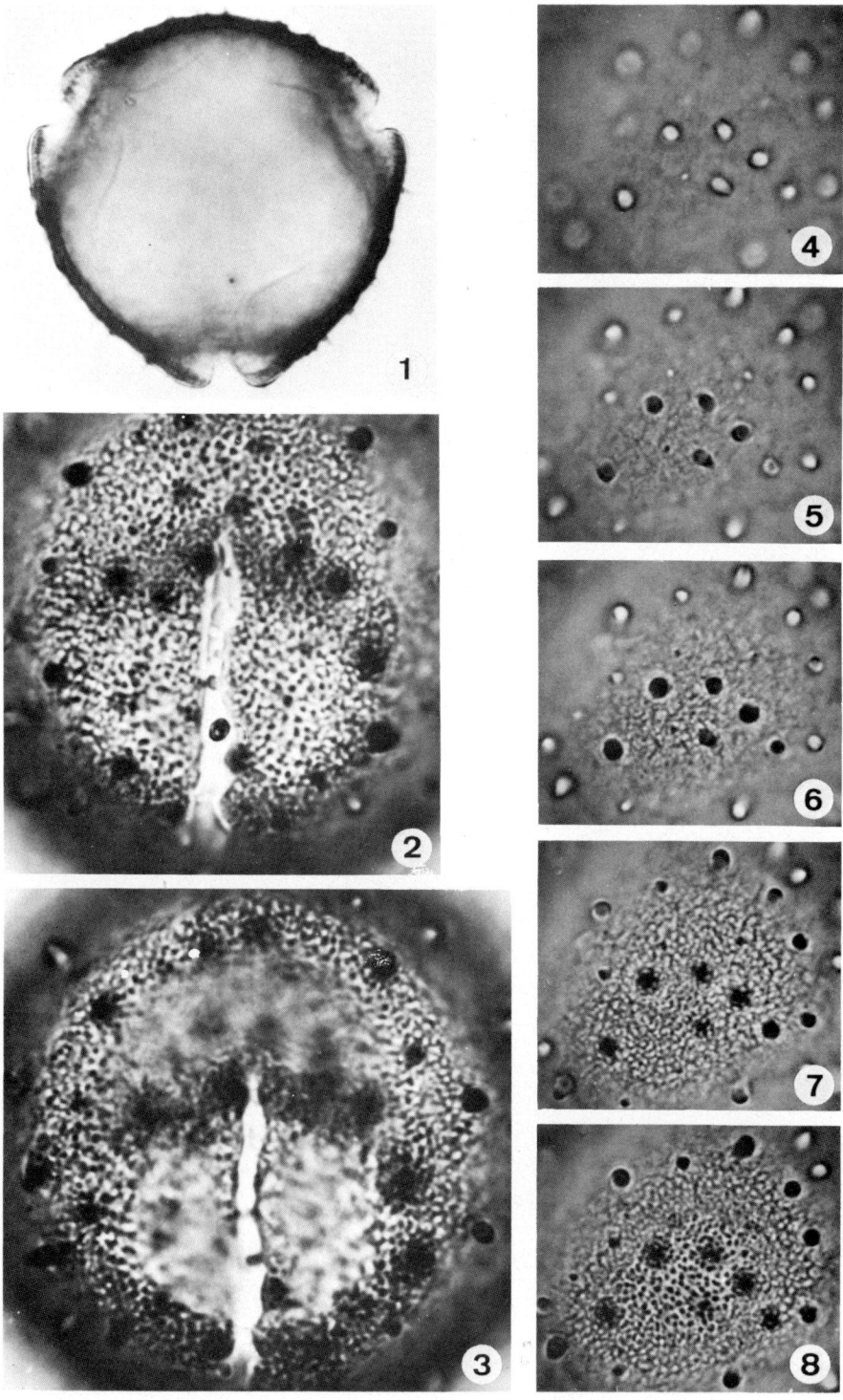

PLATE X

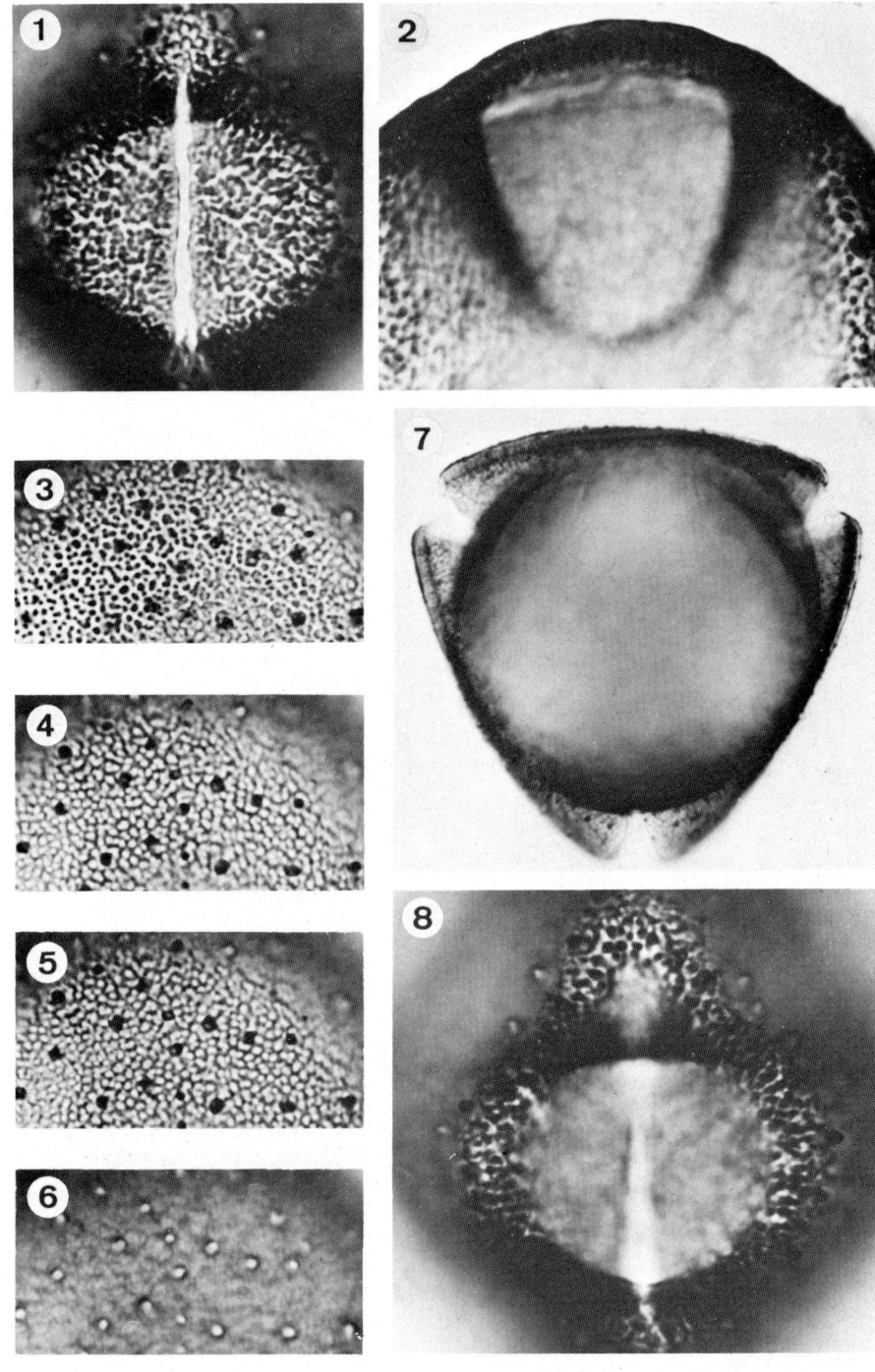

PLATE XI

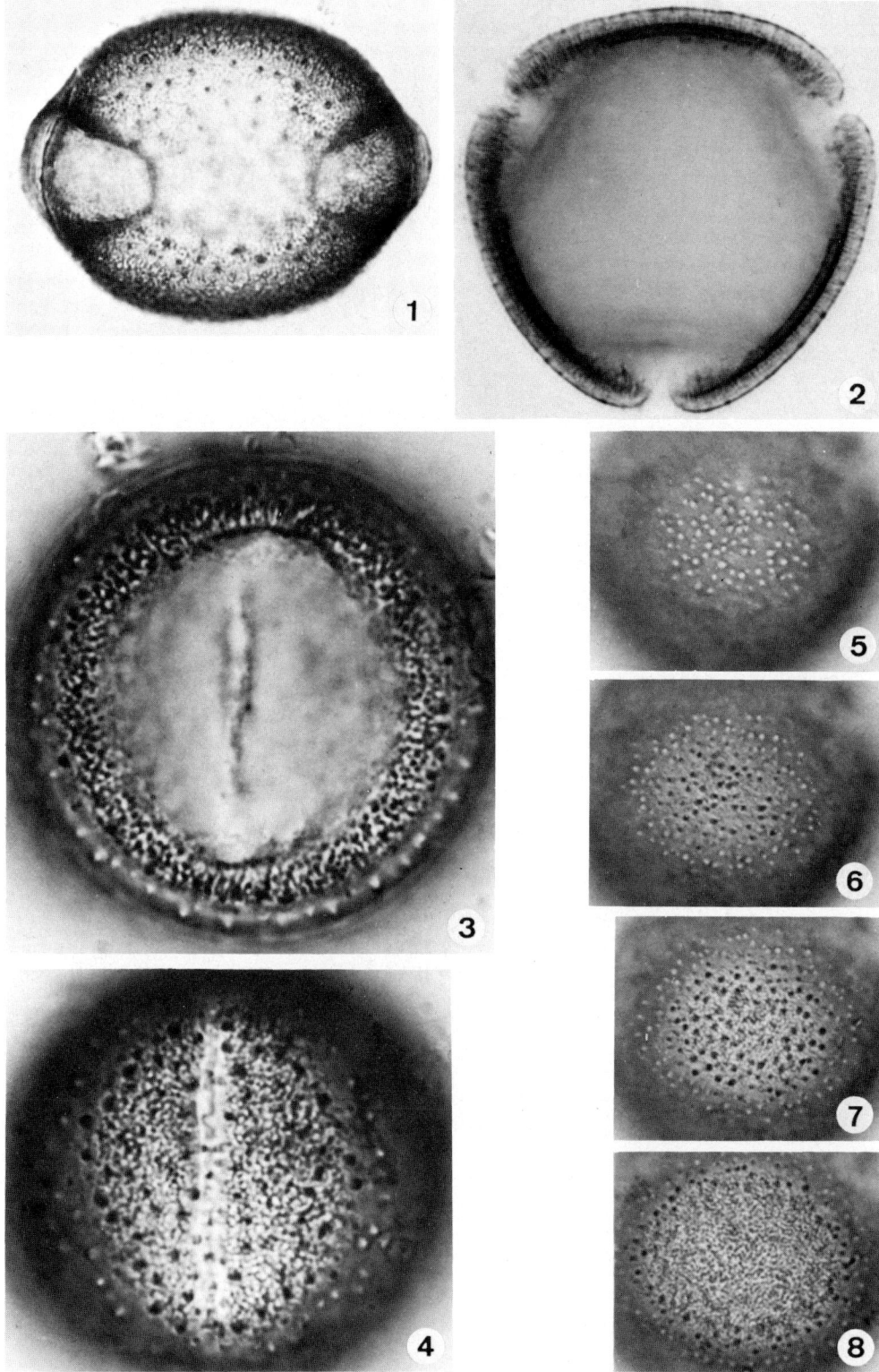

28

PLATE XII

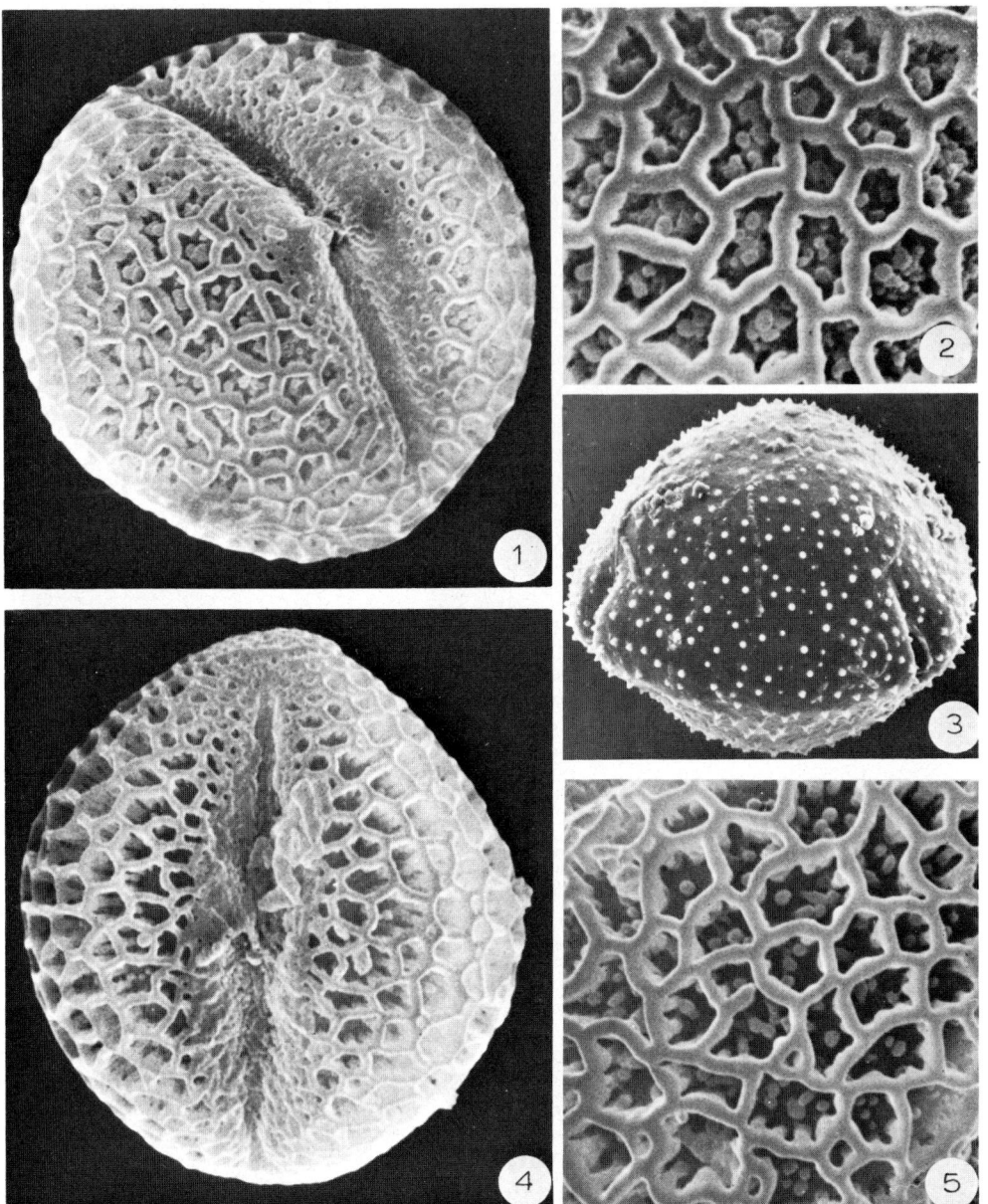

REFERENCES

Bassett, J. J. and Compton, C. W., 1970. Pollen morphology of the family Caprifoliaceae in Canada. Pollen Spores, 12; 365—380.
Erdtman, G., 1943. An Introduction to Pollen Analysis. Chronica Botanica Co., Waltham, Mass., 239 pp.
Erdtman, G., 1952. Pollen Morphology and Plant Taxonomy (An Introduction to Palynology, 1. Angiosperms). Almquist and Wiksell, Stockholm, 539 pp.
Erdtman, G., 1966. Pollen Morphology and Plant Taxonomy — Angiosperms. Hafner, New York, N.Y., 2nd ed., 553 pp.
Erdtman, G., Berglund, B. and Praglowski, J., 1961. An Introduction to a Scandinavian Pollen Flora, I. Almquist and Wiksell, Stockholm, 92 pp.
Erdtman, G., Praglowski, J. and Nilsson, S., 1963. An Introduction to a Scandinavian Pollen Flora, II. Almquist and Wiksell, Stockholm, 89 pp.
Faegri, K. and Iversen, J., 1950. Textbook of Modern Pollen Analysis. Munksgaard, Copenhagen, 169 pp.
Faegri, K. and Iversen, J., 1964. Textbook of Pollen Analysis. Munksgaard, Copenhagen, 2nd ed., 237 pp.
Kuprianova, L. A. and Alyoshina, L. A., 1972. Pollen and Spores of Plants from the Flora of the European Part of the U.S.S.R., I. Izv. Akad. Nauk S.S.S.R., Leningrad, 171 pp. (in Russian).
Radulescu, D., 1961. Palynologische Angaben betreffende die Spontaneen, sowie einige kultivierte Arten der Caprifoliaceae aus der rumänischen Volksrepublik. Lucrarille Grad. Bot. Bucuresti, 1960: 289—299.
Reitsma, Tj., 1966. Pollen morphology of some European Rosaceae. Acta Bot. Neerl., 15: 290—307.
Reitsma, Tj., 1970. Suggestions towards unification of descriptive terminology of angiosperm pollen grains. Rev. Palaeobot. Palynol., 10: 39—60.
Reitsma, Tj. and Reuvers, A. A. M. L., 1974. Northwest European pollen flora. Adoxaceae. Rev. Palaeobot. Palynol., 18 (in press).
Richard, P., 1970. Atlas pollinique des arbres et de quelques arbustes indigènes du Quebec. Nat. Can., 97: 282—306.

The Northwest European Pollen Flora, 3

PRIMULACEAE

W. PUNT, JEANETTE S. DE LEEUW VAN WEENEN AND W. A. P. VAN OOSTRUM

Laboratory of Palaeobotany and Palynology, State University, Utrecht (The Netherlands)

LITERATURE

Beug (1961), Erdtman (1952), Erdtman et al. (1961), Faegri and Iversen (1950, 1964), Hiirsalmi (1969), Huynh (1970), Spanowsky (1962), Tarnavaschi and Mitroiu (1960), Yordanov and Peev (1970), Wendelbo (1961).

TERMINOLOGY

Apocolpial field (= Apicalfeld Beug, 1961): Field situated at the poles of syncolpate pollen grains, delimited by the margins of anastomosed colpi.
NB: *Polarfeld* Beug (1961) and *Polar Area* Iversen and Troels-Smith (1950) are synonyms of *Apocolpium* Erdtman (1952).
Bridge (Faegri and Iversen, 1950, 1964, emend.): The margins of the colpi are raised in the equatorial area and connected with each other; this connection forms a bridge over the colpus, dividing the ectoaperture into two parts.
Endocingulus (Reitsma, 1966): Ring-shaped endoaperture, laying in the equatorial plane.
Endocracks (Oldfield, 1959): Grooves in the inner face of the nexine.
Horn (Huynh, 1970, emend.): Elongated part of the endoaperture, which is curved towards one of the poles (Plate I, 3).
Microreticulate (Praglowski and Punt, 1974): A network consisting of muri which encompass lumina less than 1μ in width. The breadth of the muri is equal to, or narrower than, the width of the lumina.

SPECIMENS EXAMINED

Anagallis arvensis L. ssp. *arvensis* — Germany: Biol. exc. 1965—912 (U); Ireland: Hessel, Klein and Rubers 741 (U); The Netherlands: Punt s.n., Anno 1957 (fresh material); Punt s.n., Anno 1961 (fresh material); Swart 264 (U).
Anagallis arvensis L. ssp. *coerulea* (Gouan) Vollmer — France: Leeuwenberg 1018(U); Biol. exc. 1953—814 (U); Biol. exc. 1965—2023 (U).
Anagallis crassifolia Thore — Spain: Biol. exc. 1962—1142 (U); Biol. exc. 1965—912 (U).
Anagallis foemina Miller — syn. *Anagallis arvensis* L. ssp. *coerulea* (Gouan) Vollmer.
Anagallis minima (L.) E. H. L. Krause — The Netherlands: Arendsen-Hein s.n. (U); Borrias and de Bruyn s.n. (U).
Anagallis tenella (L.) L. — Ireland: Hessel, Klein and Rubers 940 (U); The Netherlands: Mennega s.n. (U).

Androsace carnea L. — Spain: Biol. exc. 1964—1317 (U); Switzerland: Biol. exc. 1958—547 (U).
Androsace elongata L. — Czechoslovakia: Svestka 451 (U): Germany: Behrendsen s.n. (U).
Androsace lactea L. — Switzerland: Brenne s.n. (U); Biol. exc. 1965—1432 (U).
Androsace maxima L. — Austria: Kovats s.n. (U); Hungaria: Keinitzen 4197 (U).
Androsace septentrionalis L. — Germany: s.c., s.n., Herb. Utrecht No. 202205 B (U); Sweden: Minnaert s.n. (U).
Androsace villosa L. — France: Buser s.n. (U); v. Oordt s.n. (U).
Asterolinon linum-stellatum (L.) Duby — Italy: Kramer and Westra 3498 (U); Portugal: Matos and Marques s.n. (U).
Centunculus minimus L. — syn. Anagallis minima (L.) E. H. L. Krause.
Cyclamen hederifolium Aiton — France: Stefani s.n. (U); Yugoslavia: H. and P. J. M. Maas 39 (U).
Cyclamen purpurascens Miller — Austria: Schütz s.n. (U); Yugoslavia: Biol. exc. 1963—4315 (U).
Glaux maritima L. — Norway: Biol. exc. 1970—196 (U); The Netherlands: Roosma s.n. (U); Florschütz, Gadella and Mennega s.n. (U).
Hottonia palustris L. — Germany: Behrendsen s.n. (U); France: D'Alleizette s.n. (U); The Netherlands: Leeuwenberg 237 (U); Verhoeven s.n. (fresh material); s.c., s.n., Herb. Utrecht No. 009088 (U).
Lysimachia ephemerum L. — Spain: Biol. Exc. 1967—2180 (U); The Netherlands: Cult. Hort. Cantonspark 8410 (U).
Lysimachia nemorum L. — Germany: Punt s.n. (U); The Netherlands: Stafleu s.n. (U); Punt s.n. (fresh material).
Lysimachia nummularia L. — Ireland: Hessel, Klein and Rubers 1364 (U); The Netherlands: Punt s.n. (fresh material); Reitsma s.n. (fresh material); Rem s.n. (fresh material); Van der Aa s.n. (U); Yugoslavia: Biol. exc. 1962—1253 (U).
Lysimachia punctata L. — **Germany: Hinterbuber 459 (U).**
Lysimachia terrestris Britton, Sterns et Pogg. — England: Mennega s.n. (U).
Lysimachia thyrsiflora L. — Finland: Haakana s.n. (U); Norway: Biol. exc. 1970—15. (U); The Netherlands: Leeuwenberg s.n. (U); Mennega s.n. (U); Punt s.n. (fresh material).
Lysimachia vulgaris L. — Ireland: Hessel, Klein and Rubers 1504 (U); The Netherlands: Dieleman 384 (U); Punt s.n. (fresh material).
Primula elatior (L.) Hill — Belgium: Smit 2002 (U); The Netherlands: Hekking s.n. (U); Van der Burgh s.n. (fresh material).
Primula farinosa L. — England: Waid s.n. (U); Sweden: Manten s.n. (U).
Primula hirsuta Allioni — France: Biol. exc. 1960—309 (U): Italy: Behrendsen s.n. (U).
Primula scotica Hooker — England: Barkman 2392 (L); Hoogland 1948—501 (L).
Primula stricta Horneman — Sweden: Leyonsmarck s.n. (U); Wikström s.n. (U).
Primula veris L. — Ireland: Hessel, Klein and Rubers 114 (U); The Netherlands: Jonker-Verhoef and Jonker s.n. (U); Rem s.n. (fresh material); Germany: Van der Burgh s.n. (fresh material).
Primula vulgaris Hudson — Austria: Behrendsen s.n. (U); England: Howard 254 (U); Ireland: Douw and Van Embden 30 (U); The Netherlands: Rem s.n. (fresh material); Schütz s.n. (U).
Samolus valerandi L. — Sweden: Oosterveld 0267 (U); The Netherlands: Punt s.n. (fresh material); Teunissen and Wildschut s.n. (U).
Soldanella alpina L. — Italy: Behrendsen s.n. (U); Trepper s.n. (U).
Steironema ciliata (L.) Rafinesque — Cult. Hort. Utrecht s.n. (U).
Steironema lanceolata (Walter) A. Gray — Belgium: Lejeune 1010 (U).
Trientalis europaea L. — Belgium: Hekking 3956 (U); Norway: Biol. exc. 1956—80 (U); Biol. exc. 1970—1653 (U); The Netherlands: Reitsma s.n. (fresh material); s.c., s.n., Herb. Utrecht No. 99988A (U).

KEY TO THE POLLEN TYPES

1.a. Pollen grains heteropolar; 3-colporate, tectate, colpi nearly syncolpate at one pole and with a large apocolpium at the other pole..................... *Cyclamen purpurascens* type
   b. Pollen grains isopolar 2
2.a. Syncolpate..3
   b. Not syncolpate..7
3.a. Differentiation between sexine and nexine distinct; ornamentation finely reticulate or microreticulate; often more than one endoaperture per colpus........................... *Hottonia palustris* type
   b. Differentiation of exine usually indistinct; pollen grains tectate; exine thin and smooth; endoapertures absent or, if present, small and visible only with high magnification (ca. × 1000)....................4
4.a. 3 Colpi..5
   b. More than 3 colpi...6
5.a. Outline in polar view triangular; colpi narrow; apocolpial field distinct (triangular in outline); P/E ratio transverse or subtransverse .... *Primula farinosa* type
   b. Outline in polar view more or less circular; colpi rather broad; apocolpial field absent or indistinct; P/E ratio adequate to suberect .... *Soldanella alpina* type
6.a. Pollen grains symmetric, with 4 colpi (rarely 5) . . *Primula scotica* type
   b. Pollen grains asymmetric, usually with 6 colpi . . . *Primula stricta* type
7.a. Number of apertures 4—10...................................8
   b. Number of apertures 3.....................................12
8.a. Endoapertures distinct, usually with costae 9
   b. Endoapertures absent or very small and indistinct, usually with costae..11
9.a. Pantocolporate; inner face of nexine often with small processes (nexine 2); grains large (more than 30μ, up to 55μ) .... *Trientalis europaea* type
   b. Zonoaperturate; inner face of nexine smooth; grains smaller 10
10.a. Distinctly reticulate; size of lumina decreasing towards colpi.......................... *Lysimachia vulgaris* type
                                                     (*Lysimachia nummularia*)
   b. Tectate........................... *Androsace maxima* type
11.a. Zonocolporate .... *Primula veris* type
   b. Pantocolpate or pantocolporate; inner face of nexine often with small processes (nexine 2); grains large (more than 30μ, up to 55μ) .... *Trientalis europaea* type
12.a. Reticulate in mesocolpium...................................13
   b. Tectate or microreticulate 16
13.a. Endoapertures small and indistinct, without costae, often more than one per colpus .... *Hottonia palustris* type
   b. Endoapertures distinct, with costae 14

14.a. Endoapertures tapering, without horns .... *Lysimachia vulgaris* type
   b. Endoapertures with horns, at least one of the colpi with horns; sometimes endoapertures anastomosing into an endocingulus 15
15.a. P/E ratio usualy suberect, less often semi-erect; small fastigium present............................... *Anagallis arvensis* type
   b. P/E ratio usually erect, less often semi-erect; without fastigium......................... *Lysimachia ephemerum* type
16.a. Differentiation between sexine and nexine indistinct; ornamentation in surface view faint or absent 17
   b. Differentiation between sexine and nexine distinct; ornamentation in surface view usually distinct............................. 21
17.a. Endoapertures small and narrow, visible only as a nick in the colpus; pollen grains more or less adequate; outline in polar view ± circular............................ *Samolus valerandi* type
   b. Endoapertures relatively broad and not particularly small; pollen grains usually erect, at least suberect; outline in polar view ± circular or triangular 18
18.a. Pollen grains small; polar axis usually less than 20μ; outline in equatorial view usually rectangular or nearly so; exine extremely thin 19
   b. Pollen grains larger; polar axis more than 20μ; outline in equatorial view usually elliptic; exine not particularly thin............... 20
19.a. P/E ratio usually suberect, rarely semi-erect; outline in equatorial view almost rectangular or elliptic; long sides not exactly parallel........................ *Cyclamen hederifolium* type
   b. P/E ratio usually semi-erect or erect, rarely suberect; outline in equatorial view usually distinctly rectangular; long sides usually parallel............................ *Androsace elongata* type
20.a. Margins of endoapertures diffuse; shape of endoapertures indistinct; without horns........................ *Primula hirsuta* type
   b. Endoapertures distinct; usually with horns, which sometimes anastomose...................... *Lysimachia nemorum* type
21.a. Endoapertures small, narrow-looking distinct margins, often indistinct; without costae 22
   b. Endoapertures distinct; usually with costae 24
22.a. Outline in polar view circular or only slightly triangular; outline in equatorial view elliptic................. *Samolus valerandi* type
   b. Outline in polar view distinctly triangular; outline in equatorial view often slightly rhombic 23
23.a. Tectum completely closed; nexine without processes on inner face................................. *Steironema ciliata* type
   b. Tectum with numerous perforations; nexine often with small processes on inner face................. *Trientalis europaea* type
24.a. P/E ratio usually erect, rarely semi-erect; costae colpi thick; costae endocolpi present....................................... 25
   b. P/E ratio usually suberect or semi-erect, rarely erect; costae colpi present, but not always thick............................. 26

25.a. Tectum psilate, lacking ornamentation; nexine thick, often thicker than sexine.............................*Glaux maritima* type
   b. Tectum with perforations or microreticulate; nexine about as thick as sexine, but not thicker.............*Lysimachia ephemerum* type
26.a. Endoapertures elliptic, with distinct rounded ends or rectangular with obtuse angles; outline in polar view triangular..........................*Androsace maxima* type
   b. Endoapertures tapering with sharp ends, often with horns, sometimes an endocingulus; outline in polar view nearly circular 27
27.a. Endoapertures distinct, with horns; horns sometimes anastomosed into an endocingulus; polar axis rather large (over 21μ)..........28
   b. Endoapertures distinct, usually without horns, but if present small, indistinct and not anastomosing; polar axis small, usually between 16 and 20μ, sometimes up to 22μ..............*Anagallis tenella* type
28.a. Outline in equatorial view slightly rectangular to elliptic; tectum with perforations; apocolpium index usually relatively large (0.30—0.45)......................*Lysimachia nemorum* type
   b. Outline in equatorial view elliptic; ornamentation microreticulate; apocolpium index usually relatively small (0.15—0.30)........................ *Anagallis arvensis* type

DESCRIPTION OF THE POLLEN TYPES

Anagallis arvensis type (Plate I, 1—10)

Pollen class: 3-Zonocolporate or rarely 4-pantocolporate.
P/E ratio: Suberect to semi-erect.
Apertures: Ectoaperture — colpus, long, very narrow, slit-like, deeply sunken, margins distinct; ends acute; no costae or narrow, tapering ones up to the middle of the colpus; fastigium. Endoaperture — endocingulus with horns or colpus with anastomosed horns, rather broad; margins distinct; distinct costae.
Exine: Sexine about as thick as nexine or slightly thinner. Sexine 1 consisting of thin, short columellae. Sexine 2 consisting of a semi-tectate or rarely tectate layer of distinct capita.
Ornamentation: Reticulate, or microreticulate, rarely tectate. Lumina of the reticulum in the outer part of sexine 2 larger than in the inner part, irregular in shape, slightly decreasing towards the colpi, not or scarcely decreasing towards the apocolpium; capita of the muri thin in the upper part and thicker towards the lower part; simplicolumellate.
Outlines: Equatorial view — elliptic. Polar view — circular with intruding colpi.
Measurements: Glycerine jelly — P $23-35\mu$, E $20-29\mu$, P/E ratio 1.04—1.30, exine ca. 2μ thick; apocolpium index 0.15—0.30. Silicone oil — P $21-28\mu$, E $18-24\mu$ P/E ratio 1.05—1.31.

Species: Anagallis arvensis ssp. *arvensis, A. arvensis* ssp. *coerula* (syn. *Anagallis foemina*)

Comments
It was not possible to differentiate between the pollen of two subspecies of *Anagallis arvensis*. Thus the data do not provide evidence for the taxonomic decision to separate the taxa *Anagallis arvensis* and *Anagallis foemina* at the species level. This type, however, distinctly differs from the *Anagallis tenella* type by its endocingulum and the corns of the endoapertures. Otherwise the two types have many features in common.
On the other hand, the *Anagallis arvensis* type resembles the *Lysimachia nemorum* type. Their apertures and outlines are remarkably similar, the only difference being the ornamentation. The *Anagallis arvensis* type is reticulate or microreticulate, an ornamentation easily visible with a magnification of 400 ×, whereas the *Lysimachia nemorum* type is tectate without any ornamentation visible at 400 × magnification.

Anagallis tenella type (Plate II, 1—5, 9—16)

Pollen class: Usually 3-zonocolporate or rarely 4-zonocolporate.
P/E ratio: Suberect to semi-erect or rarely erect.
Apertures: Ectoaperture — lalongate colpus long, narrow, slit-like, distinctly sunken; distinct margins; margo absent or present; ends acute; with or without costae colpi; fastigium occasionally present, but if so, only small.
Endoaperture — lalongate colpus, usually tapering, sometimes with small horns; margins sharp; costae faint or distinct.
Exine: Sexine about as thick as nexine. Sexine 1 thin; consisting of short, indistinct, thin columellae. Sexine 2 a semi-tectate or tectate layer thicker than sexine 1.
Ornamentation: In mesocolpium usually microreticulate, sometimes tectate with distinct perforations; in apocolpium either a closed tectum or a tectum with many perforations. Lumina of the reticulum decreasing both towards the colpi and the apocolpium, but sometimes lumina not decreasing. Where the lumina decrease, a more or less distinct margo may be present.
Outlines: Equatorial view — elliptic. Polar view — circular with distinctly intruding colpi.
Measurements: Glycerine jelly — P 14—22μ, E 12—18μ, P/E ratio 1.01—1.43; exine ca. 1.5μ; apocolpium index 0.17—0.30. Silicone oil — P 15—23μ, E 13—17μ, P/E ratio 1.01—1.38.
Species: Anagallis crassifolia, A. minima (syn. *Centunculus minimus*), *A. tenella, Asterolinon linum-stellatum*

Key to the species or species groups
1.a. Tectum in apocolpium closed; apocolpium index ca. 0.20; margo
distinct..................................*Anagallis crassifolia*

   b. Tectum in apocolpium perforate; apocolpium index ca. 0.25; margo less distinct or absent 2
2.a. Lumina of the reticulum decreasing towards the colpi and apocolpia; lumina small, but distinct, up to 0.5μ (microreticulate)
.... *Anagallis tenella*
Anagallis minima
   b. Perforations in mesocolpium small (tectate), not decreasing towards the colpi............................ *Asterolinon linum-stellatum*

Comments
The various species belonging to the *Anagallis tenella* type are difficult to differentiate. The differentiating characters given in the key are not all sharply delimited and most features show transitions. The most constant feature is the sexine 2 in the apocolpium. In *A. crassifolia* the sexine 2 is completely closed, whereas in *Anagallis tenella*, *A. minima* and *Asterolinon linum-stellatum* the tectate layer is perforate. Beside this, the decreasing lumina in the *Anagallis* species are more obvious than the constant size of the perforations in *Asterolinon*.
The *Anagallis tenella* type is also difficult to distinguish from the *Lysimachia nemorum* type. The latter, however, differs in endoaperture features, the size of the pollen grains, and a slightly different outline in equatorial view. Particularly *Anagallis tenella* is very similar to *Lysimachia nemorum*.

Androsace elongata type (Plate III, 1—6; Plate II, 5—8)

Pollen class: 3-Zonocolporate.
P/E ratio: Erect to pererect.
Apertures: Ectoaperture — colpus long, narrow, usually slit-like, sunken; margins distinct; colpus membrane nudate; costae present or absent. Endoaperture — lalongate colpus or porus, elliptic in outline; margins clear, rarely diffuse; costae present or absent.
Exine: Thin. Differentiation between sexine and nexine not always visible in all species. If present, sexine about as thick as nexine; sexine 1 indistinct, sexine 2 a tectate layer.
Ornamentation: Psilate. Tectum sometimes with perforations.
Outlines: Equatorial view — usually rectangular, sometimes elliptic; long sides usually parallel, sometimes slightly concave or slightly convex; short sides usually distinctly convex but sometimes slightly convex; angles obtuse. Polar view — either circular with intruding colpi or 3-angular with convex sides and obtuse angles; apertures situated in the middle of the sides.
Measurements: Glycerine jelly — P $10—20\mu$, E $7—11\mu$, P/E ratio 1.33—2.11; exine thinner than 1μ; apocolpium index: 0.20—0.35. Silicone oil — P $13—18\mu$, E $7—12\mu$, P/E ratio 1.25—2.30.
Species: A. carnea, A. elongata, A. lactea, A. septentrionalis, A. villosa

Comments
The pollen grains in this type are characterized by their small size, thin exine, lack of differentiation between sexine and nexine and their P/E ratio. The elongate endoporus is, to some extent, also a differential characteristic since this feature is lacking in many primulaceous pollen types. Because of the small size of the pollen grains it is difficult to subdivide the type. However, with the aid of an oil immersion objective, it is possible to identify several species and groups.

Key to species or species groups
1.a. Endocolpus lalongate, distinct; pollen grains distinctly rectangular, short sides usually only slightly convex; exine distinctly differentiated into sexine and nexine .... *A. villosa*
  b. Endoporus lalongate, elliptical in outline; pollen grains elliptic with rounded ends; sexine and nexine not clearly differentiated 2
2.a. P/E ratio usually pererect; endoaperture and endocolpus or a rather long elongate endoporus .... *A. lactea*
  b. P/E ratio usually erect; endoaperture more distinctly a lalongate endoporus .... *A. elongata* group

The *A. elongata* group and *A. lactea* are difficult to separate, but are nonetheless quite clear. The following species belong to the *A. elongata* group: *A. elongata, A. carnea* and *A. septentrionalis.*
Wendelbo (1961) as well as Spanowsky (1962) could find no differences in the pollen grains of different species in the genus *Androsace*. Both authors, however, pointed out the pollenmorphological relationship between *Androsace* and *Cyclamen* species.

Androsace maxima type (Plate III, 7—10)

Pollen class: 3-Zonocolporate, sometimes 4-, 5- or 6-zonocolporate.
P/E ratio: Suberect.
Apertures: Ectoaperture — colpus long, narrow, sunken; margins rather vague; ends acute; without costae. Endoaperture — lalongate porus or colpus; margins clear; outline more or less rectangular with obtuse ends. The margins of the ectoaperture are arched over the endoaperture but, as the exine is not differentiated into two layers, there is no real fastigium present.
Exine: Sexine about as thick as nexine. Sexine 1 consists of indistinct, thin, short columellae. Sexine 2 is a tectate layer with perforations or grooves.
Ornamentation: The surface shows many striae or rugulae, small in size and short in length.
Outlines: Equatorial view — elliptic to slightly rhombic. Polar view — triangular with slightly convex sides and apertures situated in the obtuse angles.
Measurements: Glycerine jelly—P 21—24μ, E 19—23μ, P/E ratio 1.02—1.13; exine ca. 2μ; apocolpium index 0.20—0.30. Silicone oil—no measurements made.
Species: Androsace maxima

Comments
The *Androsace maxima* type differs considerably from the *Androsace elongata* type to which all the other *Androsace* species studied belong. Shape and size are different and also the structure of the sexine. The features of the endoaperture, on the other hand, show a close resemblance to those of the *Androsace elongata* type.

Cyclamen hederifolium type (Plate IV, 1—4)

Pollen class: 3-Zonocolporate.
P/E ratio: Suberect to semi-erect.
Apertures: Ectoaperture — colpus, long, very narrow, slit-like and slightly sunken; margins distinct; ends acute; faint costae colpi; apocolpium relatively large. Endoaperture — lalongate, relatively large and broad endocolpus; rectangular in outline with obtuse angles; margins with sharp limitations; faint costae.
Exine: Thin; sexine as thick as nexine. Sexine 1 and sexine 2 indistinctly differentiated. Sexine 2 is a closed, tectate layer.
Ornamentation: Psilate; surface smooth, without perforations.
Outlines: Equatorial view — elliptic to nearly rectangular; long sides nearly parallel; short sides distinctly convex. Polar view — circular with intruding colpi; sometimes slightly triangular with the colpi in the middle of the sides and angles obtuse.
Measurements: Glycerine jelly — P 10—15μ; E 8—11μ; P/E ratio 1.10—1.30; exine ca. 1μ thick; apocolpium index ca. 0.5. Silicone oil — P 12—14μ; E 9—12.5μ; P/E ratio 1.03—1.15.
Species: Cyclamen hederifolium

Comments
In many respects the *Cyclamen hederifolium* type resembles the *Androsace elongata* type. Corresponding characters include an indistinct differentiation of the sexine, a completely closed tectum and, especially, the outline in equatorial view. In the *Cyclamen hederifolium* type, however, the long sides of the outline in equatorial view are not exactly parallel: the outline as a whole is elliptic, whereas the *Androsace elongata* type shows an outline which is more exactly rectangular with parallel long sides.
The *Cyclamen hederifolium* type is clearly distinguishable from the *Cyclamen purpurascens* type by its isopolar apocolpia and different P/E ratio.

Cyclamen purpurascens type (Plate III, 11—15)

Pollen class: Pollen grains heteropolar; at one pole, colpi syncolpate or nearly so, at the other pole with a large apocolpium; 3-zonocolporate.
P/E ratio: Adequate or subtransverse.

Apertures: Ectoaperture — colpus long, very narrow, slit-like; slightly sunken; margins distinct; ends acute; costae colpi rather broad and with sharp limitations. Endoaperture — lalongate colpus; margins sharp; outline rectangular with obtuse angles; costae.
Exine: Differentiation between sexine and nexine indistinct. Sexine 2 is a tectate layer.
Ornamentation: Psilate. Surface smooth, without perforations.
Outlines: Equatorial view — bilaterally symmetric; ovate with the largest breadth slightly shifted towards the large apocolpium. Polar view — more or less 3-angular, with slightly convex sides, and apertures situated in the obtuse angles.
Measurements: Glycerine jelly — P 12—17μ, E 12—18μ, P/E ratio 1.00—0.93; exine ca. 1μ thick; apocolpium index at one pole 0.00 (syncolpate)—0.10; at the other pole larger 0.50. Silicone oil — P 12—14μ, E 12—14μ, P/E ratio 1.00—0.92.
Species: C. purpurascens

Comments
The *Cyclamen purpurascens* type differs from the *C. hederifolia* type in its heteropolarity, ovate shape and P/E ratio, which is adequate or slightly subtransverse. These characteristics also distinguish the type from all other primulaceous pollen types studied.

Glaux maritima type (Plate IV, 5—10; Plate XIV, 1—2)

Pollen class: 3-Zonocolporate.
P/E ration: Semi-erect to erect.
Apertures: Ectoaperture — colpus long, very narrow, slit-like and deeply sunken; distinct margins; ends acute; costae colpi thick; usually a small fastigium present. Endoaperture — lalongate colpus; usually elliptic in outline, with obtuse ends, often of varying number of horns; margins distinct but often slightly broken up by numerous granules; costae.
Exine: Exine rather thick. Sexine about as thick as nexine in the apocolpium, but distinctly thinner in the mesocolpium. Sexine 1 thinner than sexine 2, sexine 1 consisting of thin, numerous columellae; sexine 2 is a tectate layer.
Ornamentation: Psilate. Tectum with many small perforations in the mesocolpium; perforations fewer or absent in the apocolpium.
Outlines: Equatorial view — elliptic. Polar view — circular with deeply intruding colpi.
Measurements: Glycerine jelly — P 23—31μ, E 17—20μ, P/E ratio 1.32—1.70; exine ca. 2μ thick; apocolpium index 0.16—0.30, but usually ca. 0.20. Silicone oil — P 23—26μ, E 18—19μ, P/E ratio 1.29—1.40.
Species: Glaux maritima

Hottonia palustris type (Plate IV, 11—12; Plate V, 1—6; Plate XIV, 6)

Pollen class: 3-Zonocolporate; rarely syncolpate or 2-zonocolporate.
P/E ratio: Semi-erect to subtransverse, but usually suberect.
Apertures: Ectoaperture — colpus long, narrow, but not very narrow, deeply sunken; margins distinct; ends acute; no costae; fastigium absent or faintly present. Endoaperture — lalongate colpus or porus; indistinct margins; no costae. In pollen grains of brachystylous specimens the number of endoapertures is often more than one, sometimes up to three. Apocolpium small, varying in size.
Exine: Thin. Sexine thicker than nexine. Sexine 1 thinner than sexine 2; sexine 1 consisting of columellae. Sexine 2 is a semi-tectate layer.
Ornamentation: Pollen grains reticulate or microreticulate. Muri thin, and always narrower than the width of the lumina. In brachystylous specimens the pollen grains are reticulate and the lumina decrease towards the apocolpium where the reticulum is finer (microreticulate). In dolichostylous specimens the ornamentation of the pollen grains is microreticulate. Lumina irregular in shape.
Outlines: Equatorial view — circular to elliptic. Polar view — circular with deeply intruding colpi.
Measurements: Brachystylous specimens: glycerine jelly — P 18—24μ, E 16—20μ, P/E ratio 1.05—1.31. Dolichostylous specimens: glycerine jelly — P 10—13μ, E 10—13μ, P/E ratio 0.95—1.10; silicone oil — P 12—15μ, E 12—13μ, P/E ratio 0.94—1.12.
Exine ca. 1μ thick; lumina of the reticulum ca. 1μ in diameter, up to 1.5μ in pollen of brachystylous specimens; apocolpium index usually between 0.10 and 0.20, rarely up to 0.00 (syncolpate).
Species: Hottonia palustris

Comments
Hottonia is one of the genera with dimorphic pollen and as in the genus *Primula* this dimorphism is connected with heterostyly. The differences between the pollen grains of brachystylous and dolichostylous flowers are mainly based on size and shape. The only additional differing feature is that in the pollen grains of dolichostylous specimens the number of endoapertures does not vary; all specimens having only one endoaperture. Otherwise the pollen grains are smaller and have a more adequate P/E ratio.
The type is characterized by the very long colpi, indistinct endoapertures and microreticulate ornamentation.
Erdtman et al. (1961) and Wendelbo (1961) give essentially the same description and sizes of *Hottonia* pollen grains.

Lysimachia ephemerum type (Plate VIII, 5—9)

Pollen class: 3-Colporate.
P/E ratio: Erect to semi-erect.
Apertures: Ectoaperture — colpus, long, very narrow, slit-like, sunken; margins distinct; ends acute; costae colpi tapering towards the poles; no fastigium.

Endoaperture — colpus lalongate; margins with sharp limitations; costae; horns, one or two per colpus end, sometimes anastomosing into an endocingulum.
Exine: Sexine about as thick as nexine or slightly thinner. Sexine 1 slightly thinner than or as thick as sexine 2. Sexine 1 consisting of thin, widely spaced columellae; sexine 2 is a semi-tectate layer.
Ornamentation: Reticulate or microreticulate. Lumina usually wider than muri; angular in shape; distinctly decreasing towards the poles and only slightly decreasing towards the colpi. Muri thin, simplicolumellate.
Outlines: Equatorial view — elliptic, long sides slightly emarginate at the equator. Polar view — circular to slightly angular; with convex sides and apertures situated in the obtuse angles.
Measurements: Glycerine jelly — P 24—31μ, E 18—21μ, P/E ratio 1.20—1.53; Silicone oil — P 23—29μ, E 18—21μ, P/E ratio 1.25—1.41.
Exine varying from 1.5 to 2μ; lumina usually about 1μ wide, sometimes less than 1μ; apocolpium index 0.25—0.45.
Species: Lysimachia ephemerum

Comments
The *Lysimachia ephemerum* type is characterized by its P/E ratio, distinct endoapertures with horns and decreasing lumina towards the poles. In the last two features the type has some affinities with the *Anagallis* types. The erect P/E ratio, however, keeps the present type apart from the semi-erect to suberect *Anagallis* types. The type also resembles the *Lysimachia vulgaris* type, but can be distinguished by the three features mentioned above.

Lysimachia nemorum type (Plate VIII, 10; Plate IX, 8—11; Plate XIV, 4)

Pollen class: 3-Zonocolporate.
P/E ratio: Suberect to semi-erect, rarely erect.
Apertures: Ectoaperture — colpus long, very narrow, slit-like, deeply sunken; distinct margins; acute ends; narrow costae, tapering up to the middle of the colpus; fastigium absent or small and indistinct. Endoaperture — lalongate colpus, large and relatively broad, with one or more distinct horns, sometimes, but not always anastomosed into an endocingulus; costae.
Exine: Sexine about as thick as nexine. Sexine 1 thinner than sexine 2; sexine 1 consisting of thin columellae. Sexine 2 is a tectate layer.
Ornamentation: The tectum is distinctly perforated with a tendency towards a microreticulum.
Outlines: Equatorial view — more or less rectangular or elliptic. Polar view — circular with intruding colpi.
Measurements: Glycerine jelly — P 21—24μ, E 17—21μ, P/E ratio 1.05—1.37; exine ca. 1μ up to 1.5μ; apocolpium index 0.30—0.45. Silicone oil — P 18—22μ, E 16—19μ, P/E ratio 1.05—1.22.
Species: Lysimachia nemorum

Comments

The *Lysimachia nemorum* type is easily differentiated from the other *Lysimachia* types by its tectate sexine. The *Steironema ciliata* type also shows a tectate sexine, but it is triangular in polar view, whereas the *Lysimachia nemorum* type is circular in polar view.

The *Lysimachia nemorum* type can only be distinguished with difficulty from the *Anagallis arvensis* type. The pollen grains of the latter type are slightly larger; they have a more distinct ornamentation and a more elliptic outline in equatorial view. It is the combination of differential characters which separates the two types.

Lysimachia vulgaris type (Plate VI, 1—8; Plate VII, 1—9; Plate VIII, 1—4; Plate IX, 1—7)

Pollen class: 3-(4-)Zonocolporate.
P/E ratio: Semi-erect to suberect.
Apertures: Ectoaperture — colpi, long, very narrow, slit-like, sunken; margins distinct; margo; ends acute; costae colpi tapering from the equatorial plane towards the poles; fastigium sometimes present, usually absent.
Endoaperture — lalongate colpus, usually tapering, with distinct, acute ends; sometimes ends diffuse, rarely with indistinct horns; costae tapering from the ectocolpus towards the outer ends.
Exine: Sexine about as thick as nexine. Sexine 1 distinct, about as thick as sexine 2 or slightly thinner; sexine 1 thin, consisting of columellae. Sexine 2 is a semi-tectate layer.
Ornamentation: Reticulate. Lumina irregular, angular; distinctly decreasing towards the colpi, thus forming a margo. Muri usually simplicolumellate, rarely duplicolumellate, in the mesocolpium narrower than the lumina diameter and broader towards the colpi. In surface view the outline of the collumellae is circular.
N.B.: In *L. punctata* the lumina are provided with small scattered granules (scabrae).
Outlines: Equatorial view — elliptic to slightly rectangular, with convex sides and obtuse ends. Polar view — circular to slightly angular, with slightly convex sides to distinctly convex sides and obtuse angles; colpi situated in the middle of the sides.
Measurements: Glycerine jelly — P 19—33μ, E 17—28μ, P/E ratio 1.02—1.33; exine varying from 1 to 2μ; apocolpium index varying from 0.10 to 0.50; lumina of the reticulum varying from 1 to 2.5μ. Silicone oil — P 19—28μ, E 16—26μ, P/E ratio 1.01—1.26.
Species: Lysimachia nummularia, L. terrestris, L. thyrsiflora, L. punctata, L. vulgaris

Comments
The pollen grains belonging to the *Lysimachia vulgaris* type are all very much alike. Nevertheless, the type can be subdivided into three more or less distinct groups.

Key to the species or species groups
1.a. Lumina of the reticulum provided with small granulaes
   (scabrae)........................... *Lysimachia punctata*
  b. Lumina of the reticulum without granules.................. 2
2.a. Pollen grains relatively small, P (longest axis) not exceeding 25μ; a
   small, but usually distinct fastigium present...... *L. thyrsiflora* group
                         (species: *L. thyrsiflora, L. terrestris*)
  b. Pollen grains relatively large, P (longest axis) in glycerine jelly usually
   larger than 25μ and in silicone oil usually larger than 23μ; fastigium
   usually absent, sometimes indistinct .... *L. vulgaris* group
                         (species: *L. nummularia, L. vulgaris*)

Lysimachia terrestris is an introduced species from North America and its distribution in North-West Europe is scattered.
The pollen grains of the *Lysimachia vulgaris* type are related to both the *Anagallis tenella* type and *Anagallis arvensis* type. In particular the endoapertures of the *Anagallis tenella* type closely resemble those of the *Lysimachia vulgaris* type although the ornamentation is microreticulate. On the other hand, the *Anagallis arvensis* type shows a reticulum which is much like the reticulum of the *Lysimachia vulgaris* type. Both types differ distinctly in the shape of their endoapertures, the latter with horns indistinct or absent and the former with distinct, often anastomosed horns.
The *Lysimachia ephemerum* type is also related, but it differs in its P/E ratio which is clearly erect, whereas the *Lysimachia vulgaris* type is usually semi-erect to suberect, and also in its endoapertures.

Primula farinosa type (Plate V, 8—14)

Pollen class: Syncolpate; zonoaperturate, 3-zonocolpate, rarely 4-zonocolpate.
P/E ratio: Semitransverse to transverse.
Apertures: Ectoaperture — colpus, very narrow, slit-like, sunken; distinct margins; no costae; colpi closely approaching each other, but not meeting in the apocolpium; forming a distinct triangular apocolpial field. Endoaperture — absent.
Exine: Thin. Sexine about as thick as nexine. Sexine 1 and sexine 2 indistinct. Sexine 2 is a tectate layer.
Ornamentation: Psilate. Tectum sometimes perforate.
Outlines: Equatorial view — elliptic, long sides distinctly concave because of the sunken colpi at the poles; pear-shaped, if one of the colpi is situated in the plane of projection. Polar view — triangular with straight to slightly convex sides, apertures situated in the obtuse angles.
Measurements: Brachystylous specimens: glycerine jelly — P $8-11\mu$, E $12-16\mu$, P/E ratio 0.53—0.71; silicone oil — P $7-10\mu$, E $13-15\mu$, P/E ratio 0.57—0.64; exine ca. 1μ thick.
Dolichostylous specimens: glycerine jelly — P $6-8\mu$, E $8-11\mu$, P/E ratio

0.69—0.83; silicone oil — P 7—9µ, E 9—10µ, P/E ratio 0.75—0.85.
Species: Primula farinosa

Comments
The *P. farinosa* type is clearly related to the *P. scotica* type and the *P. stricta* type, the number of colpi being the main differentiating character between the three types. According to Spanowsky (1962), Wendelbo (1961), Yordanov and Peev (1970), etc., several more species in the genus *Primula* and also in some other genera (e.g. *Cortusa*) show pollen grains of the *P. farinosa* type. Those species, however, do not occur in the area under consideration.
Pollen of the *P. farinosa* type is distinctly dimorphic. The pollen grains of brachystylous specimens are larger in overall size and also the P/E ratio is slightly different. However, all other features of both brachystylous specimens as well as dolichostylous specimens are the same.

Primula hirsuta type (Plate X, 8—11)

Pollen class: 3-Zonocolporate.
P/E ratio: Semi-erect to erect.
Apertures: Ectoaperture — colpus long, very narrow, slit-like, deeply sunken; margins distinct; ends acute; no costae. Endoaperture — indistinct, margins diffuse, outline probably rather large but difficult to measure; no costae.
Exine: Thin. Sexine about as thick as nexine. Sexine 1 and sexine 2 indistinct. Sexine 2 is a tectate layer.
Ornamentation: Tectum psilate, with many perforations; perforations in apocolpium regular, more or less circular in outline but in the mesocolpium irregular in outline.
Outlines: Equatorial view — elliptic, short sides slightly convex, giving the poles a blunt obtuse, appearance. Polar view — triangular with slightly convex sides and the apertures situated in the obtuse angles.
Measurements: Glycerine jelly — P 18—23µ, E 15—18µ, P/E ratio 1.19—1.40; exine ca. 1µ thick; apocolpium index 0.20—0.30. Silicone oil — P 19—24µ, E 15—19µ, P/E ratio 1.17—1.32.
Species: Primula hirsuta

Comments
The *Primula hirsuta* type has no close resemblance to any other *Primula* type described in this paper. The P/E ratio together with the relatively large, but indistinct endoapertures and the tectum perforatum characterize the type. The pollen is probably dimorphic, but this was not observed in the specimens investigated.

Primula scotica type (Plate XI, 1—3)

Pollen class: Syncolpate; 4-(5-)pantocolpate; rarely zonocolpate.

P/E ratio: If zonocolpate, more or less adequate.
Apertures: Ectoaperture — colpus very narrow, slit-like; sunken; distinct margins; no costae. Colpi syncolpate at the poles or, sometimes, leaving a small, indistinct apocolpial field. Endoaperture — absent or very small and indistinct, visible only with high magnifications ($\times$ 1000).
Exine: Thin. Sexine about as thick as nexine. Sexine 1 about as thick as sexine 2. Sexine 1 consisting of indistinct, small and thin columellae; sexine 2 is a tectate layer.
Ornamentation: Psilate; tectum with many perforations.
Outlines: Usually irregular in outline, depending on the situation of the colpi. Zonocolpate pollen grains: equatorial view — circular, elliptic to more or less angular; polar view — rectangular with convex sides and colpi situated in the obtuse angles.
Measurements: Longest axis in glycerine jelly 15—18μ; in silicone oil 15—18μ; exine ca. 1μ thick.
Species: Primula scotica

Comments
The *P. scotica* type is distinctly related both to the *P. farinosa* type and to the *P. stricta* type. The distinctive features with the *P. farinosa* type are clear and mainly found in the number of colpi and regularity of the pollen grains. The absence of a distinct apocolpial field is a less exclusive distinguishing character. Pollen grains of the *P. stricta* type may occasionally be found which lack this field and, conversely, many grains of the *P. scotica* type show a distinct apocolpial field. The presence of a rudimentary endoaperture is also important in the *P. scotica* type, whereas this feature is completely absent in the *P. stricta* type. Because of these three differential characters a separation of the *P. scotica* type from the *P. stricta* type seems logical. Heterostyly does not occur in this species, so dimorphic pollen is not to be expected.

Primula stricta type (Plate V, 7, 15—16)

Pollen class: Syncolpate; (5-) 6- (7-)pantocolpate.
Apertures: Ectoaperture — colpus, very narrow, slit-like, sunken; distinct margins; no costae. Endoaperture — absent. At the apocolpium the colpi do not always meet at the poles, they often leave an apocolpial field of irregular shape with distinct margins. The shape can be triangular, rectangular, elliptic, circular, etc.
Exine: Thin. Sexine about as thick as nexine. Sexine 1, about as thick as sexine 2, consisting of indistinct columellae. Sexine 2 is a semi-tectate layer.
Ornamentation: Microreticulate merging into a tectum perforatum. Lumina extremely fine visible only with high magnification ($\times$ 1000). Sometimes lumina about as wide as muri, and the sexine 2 consequently is a perforated tectum. Tectum psilate.

Outlines: Outline irregular; as the colpi are sunken and variable in length the outline is irregularly lobed.
Measurements: Longest axis in glycerine jelly 18—21μ; in silicone oil 19—23μ exine ca. 1μ thick.
Species: Primula stricta

Comments
The *Primula stricta* type is undoubtedly related to both the *P. farinosa* type and the *P. scotica* type. The main difference is found in the number of colpi and the irregular outline.

Primula veris type (Plate IX, 1—7; Plate X, 1—7; Plate XIV, 3)

Pollen class: Zonocolpate; number of colpi varying from (5)—6 to 9—(10).
P/E ratio: Subtransverse to suberect; rarely semi-erect or semitransverse.
Apertures: Ectoaperture — colpi long, very narrow, slit-like, usually more or less sunken; margins usually distinct, ends acute; no costae. Endoaperture — absent.
Exine: Nexine thin, sexine distinctly thicker than nexine. Sexine 1 consists of distinct columellae about as thick as sexine 2 or slightly thicker. Sexine 2 is a semi-tectate layer, in cross-section consisting of more or less circular capita. Endocracks often present in the nexine.
Ornamentation: Reticulate or microreticulate. In dolichostylous specimens the grains are usually microreticulate; muri thin, lumina more or less circular in outline, less often slightly angular. In brachystylous specimens the grains are usually distinctly reticulate; muri thin, lumina angular in outline; simplicolumellate.
Outlines: Equatorial view — circular or elliptic. Polar view — circular, either with slightly intruding colpi or colpi not intruding.
Measurements: Brachystylous specimens: glycerine jelly — P 19—27μ, E 22—27μ, P/E ratio 0.78—1.18; silicone oil — P 19—25μ, E 20—26μ, P/E ratio 0.89—1.01.
Dolichostylous specimens: glycerine jelly — P 12—20μ, E 12—16μ, P/E ratio 0.80—1.18; silicone oil — P 15—17μ, E 14—16μ, P/E ratio 0.90—1.13.
Exine in brachystylous specimens ca. 1.5μ; in dolichostylous specimens ca. 1μ. Lumina of the reticulum in brachystylous specimens usually ca. 1μ wide, sometimes up to 1.5μ. Lumina in dolichostylous specimens less than 1μ wide (microreticulate).
Species: Primula elatior, P. veris, P. vulgaris

Key to the species (brachystylous specimens)
1.a. Pollen grains adequate to suberect, rarely semi-erect; colpi with distinct margins, more or less sunken .... *P. vulgaris*
   b. Pollen grains semitransverse to subtransverse, rarely adequate; colpi usually not or only slightly sunken, with distinct or indistinct margins. 2

2.a. Colpi distinct, usually slightly sunken, rarely not sunken, margins
distinct, straight . . . . *P. elatior*
b. Colpi rather indistinct, not sunken, rarely slightly sunken, margins
less clear . . . . *P. veris*

Comments
All three species belonging to this type have dimorphic pollen. The main difference between the pollen grains of brachystylous specimens and dolichostylous ones is in size, but the ornamentation also differs. In brachystylous specimens the pollen grains are reticulate, i.e. the lumina are more than 1μ wide and their outline is irregular and angular. The dolichostylous ones have lumina less than 1μ wide (microreticulate) and their outline is more or less circular and less angular.

The most characteristic features of the types are the absence of endoapertures and the colpus number which is more than 4. Another important feature is the thin nexine. Because of the latter feature the outline of the pollen grains is rather variable and less reliable for classification.

Despite this difficulty the authors have attempted to produce a key to the species. This key, however, is based on brachystylous specimens alone as it is extremely difficult to classify the dolichostylous specimens where the differential characters are less clear.

The definition which Spanowsky (1962) gives of his *veris*-type is certainly less detailed than that of the present *Primula veris* type. It comprises among others pollen grains with three colpi. On the other hand, he lays much emphasis on the fact that the colpi are not sunken. According to the present observations, however, several specimens belonging to this type, are provided with more or less sunken colpi.

Samolus valerandi type (Plate XI, 4—6; Plate XIII, 9—10; Plate XIV, 5)

Pollen class: 3-Zonocolporate.
P/E ratio: Suberect to semi-erect; rarely erect.
Apertures: Ectoaperture — colpus long, narrow, deeply sunken; margins distinct; ends acute; no costae; apocolpium relatively small; fastigium small, but distinct. Endoaperture — indistinct, probably elliptic, lalongate; margins diffuse; no costae. Endoaperture only visible as a knick in the colpus.
Exine: Thin. Sexine about as thick as nexine or slightly thicker. Sexine 1 consisting of indistinct thin columellae; sexine 2 is a tectate layer.
Ornamentation: Psilate. Tectum with many small perforations, present in mesocolpium as well as apocolpium.
Outlines: Equatorial view — elliptic. Polar view — circular with intruding colpi.
Measurements: Glycerine jelly — P $16-19\mu$, E $12-17\mu$, P/E ratio 1.05—1.25; exine ca. 1μ; apocolpium index 0.15—0.25. Silicone oil — P $14-17\mu$, E $13-16\mu$, P/E ratio 1.06—1.11.
Species: Samolus valerandi

Soldanella alpina type (Plate XIII, 1—3, 7—8)

Pollen class: Syncolpate; number of colpi 3, rarely 4.
P/E ratio: Suberect to subtansverse, rarely semi-erect.
Apertures: Ectoaperture — colpus rather broad at the apocolpium, and narrow towards the mesocolpium, slightly sunken; margins diffuse; colpus membrane granulate; equatorial bridge usually present; no costae. Endoaperture — absent.
Exine: Thin. Sexine thicker than nexine; sexine 1 consisting of indistinct columellae; sexine 2 is a tectate layer.
Ornamentation: The imperforate tectum is covered with small, granular scabrae.
Outlines: Equatorial view — circular to slightly elliptic. Polar view — circular with deeply intruding colpi.
Measurements: Glycerine jelly — P 18—21μ, E 17—22μ, P/E ratio 0.95—1.17; exine ca. 1μ thick. Silicone oil — P 16—21μ, E 16—21μ, P/E ratio 0.97—1.03.
Species: Soldanella alpina

Steironema ciliata type (Plate XI, 7—11)

Pollen class: 3-Zonocolporate; rarely 4-zonocolporate and than 4-loxocolporate.
P/E ratio: Usually suberect to semi-erect; rarely adequate or subtransverse.
Apertures: Ectoaperture — colpus long, narrow and sunken; margins distinct; ends acute; no costae; small fastigium present. Endoaperture — lalongate colpus, indistinct, very small and narrow; margins diffuse; small, but distinct costae, tapering towards the ectocolpi.
Exine: Thin. Sexine about as thick as nexine. Sexine 1 about as thick as sexine 2. Sexine 1 consisting of indistinct columellae; sexine 2 is a closed tectate layer.
Ornamentation: Psilate. Tectum without perforations.
Outlines: Equatorial view — outline more or less rhombic if colpi not in the plane of projection; otherwise either elliptic, or rarely, circular. Polar view — triangular with straight to slightly convex sides and apertures situated in the obtuse angles.
Measurements: Glycerine jelly — P 23—26μ, E 19—25μ, P/E ratio 0.96—1.28; exine varying from 1 to 1.6μ; apocolpium index 0.14—0.29. Silicone oil — P 23—27μ, E 23—26μ, P/E ratio 0.97—1.04.
Species: Steironema ciliata, S. lanceolatum

Comments
The *Steironema ciliata* type is characterized by the thin, tectate exine and the triangular outline in polar view. The pollen grains differ considerably from those occuring in *Lysimachia* s.str.
This additional morphological feature provides further justification for the taxonomic separation of *Steironema* and *Lysimachia* at the generic level.

Trientalis europaea type (Plate XII, 1—5; Plate XIII, 4—6)

Pollen class: Pollen dimorphic: A: 3-zonocolporate, rarely 4-zonocolporate; or B: 4—8-pantocolpate or pantocolporate.
P/E ratio: A: Subtransverse; B: no P/E ratio, pollen grains are asymmetric.
Apertures: Ectoaperture — A: colpus, long, narrow, slightly sunken; margins distinct; colpus membrane nudate; ends acute; apocolpium index relatively small. B: colpus of medium length, rather broad, slightly sunken; margins indistinct; ends rounded; colpus membrane granulate; no costae. Endoaperture — A: indistinct, small and narrow; margins indistinct; no costae. B: indistinct or absent.
Exine: Sexine about as thick as nexine or slightly thicker. Sexine 1 consisting of short, thin columellae; sexine 2 is a semi-tectate or tectate layer in optical cross-section consisting of circular or slightly elongate capita. Nexine 2 usually provided with many small, irregular, randomly arranged warts towards the inner face of the exine; these warts are inordinately placed, and irregular in shape and size. In the small grains these warts are often absent.
Ornamentation: Microreticulate to tectum perforate.
Outlines: Equatorial view — in zonocolporate grains elliptic. The pollen grains with more than 3 colpi are irregular and angular in shape. Polar view — in zonocolporate grains distinctly triangular with straight to very slightly convex sides and obtuse angles; apertures situated in the angles.
Measurements: A: Small grains, usually 3-colporate, rarely 4-colporate. In glycerine jelly — P 22—28μ, E 25—29μ, P/E ratio 0.89—0.98; exine 1.5—2μ thick. Lumina of the reticulum smaller than 1μ, apocolpium index in 3-colporate grains 0.15—0.20. In silicone oil — P 19—23μ, E 20—26μ, P/E ratio 0.89—0.96.
B: Large grains, usually with more than 3 apertures. Longest axis 25μ—50μ.
Species: Trientalis europaea

Comments
Pollen grains of *Trientalis europaea* L. are distinctly dimorphic and the two forms of pollen grains have been called group A and group B. The delimitations between these groups are, however, not sharp and many transitional forms exist. The A group with small pollen grains is usually zonocolporate with narrow colpi and small, but distinct endoapertures. The B group comprises larger pollen grains usually with more than 3 colpi in a pantocolpate position. The colpi are broader, less sunken and show no distinct endoapertures. Moreover, the nexine in the B group shows numerous warts on the inner face of the exine. These warts are usually, but not always, absent in the smaller grains of group A. Consequently pollen grains provided with these special nexine warts have a nexine which can be divided into two layers: nexine 1 and nexine 2. Nexine 1 is a united homogeneous layer next to the sexine; nexine 2 consists of warts.
The dimorphism in *Trientalis* pollen is of a different type to that in *Primula* and *Hottonia*. In the latter two genera the dimorphism is connected with

heterostyly (see p.*41*) and the pollen grains differ in size and shape only. In *Trientalis*, however, it is connected with a difference in chromosome numbers, the individuals with small pollen grains being diploid ($2n = 160$; Hiirsalmi, 1969), whereas large pollen grains are often produced by polyploid individuals. Moreover, the differences between the pollen grains of group A and group B are not only based on size and shape, but also on structural features.

According to Hiirsalmi (1969) the size of the pollen grains is to some extent dependent on external factors. Pollen grains from the same individual vary in size from year to year. This is another strong indication that measurements mentioned in pollen descriptions are often of dubious significance.

PLATE DESCRIPTIONS (all Plates × 1500, **except Plate XIV**)

PLATE I (p.*56*)

Anagallis arvensis L. ssp. *arvensis* (Hessel, Klein and Rubers 741)
1. Outline in polar view.
2. Outline in equatorial view.
7. Apocolpium.
8. Reticulum at high focus.
9. Reticulum at low focus.
10. Columellae layer in surface view.

Anagallis arvensis L. ssp. *coerulea* (Gouan) Vollm. (Biol. exc. 1965—1023)
3. Endocingulus with horns.
4. Narrow ectocolpus and broad endoaperture.
5. Apocolpium.
6. Endocingulus with anastomosing horns.

PLATE II (p.*57*)

Anagallis tenella (L.) L. (Mennega s.n.)
1. Outline in equatorial view.
2. Outline in polar view.
3. Ectocolpus without margo.
10. Reticulum in mesocolpium.
16. Reticulum in apocolpium.

Anagallis crassifolia Thore (Biol. exc. 1962—1142)
4. Ectocolpus with distinct margo, tapering endoaperture.

Androsace lactea L. (Biol. exc. 1965—1432)
5. Endoaperture.
6. Outline in equatorial view, distinct costae.

Androsace carnea L. (Biol. exc. 1964—1137)
7. Endoaperture.
8. Outline in equatorial view.

Asterolinon linum-stellatum (L.) Duby (Matos and Marques s.n.)
9. Outline in equatorial view.
13. Ectocolpus without margo, tapering endoaperture.
14. Microreticulum not decreasing towards the apocolpium.

Anagallis minima (L.) E.H.L. Krause (Arendsen-Hein s.n.)
11. Reticulum.

Anagallis crassifolia Thore (Biol. exc. 1957—912)
12. Reticulum in mesocolpium.
15. Apocolpium with closed tectum.

PLATE III (p.58)

Androsace villosa L. (Buser s.n.)
1. Outline in equatorial view.
2. Outline in polar view.
3. Elongated endoaperture and distinct costae.
4. Slit-like ectocolpus.
5. Ornamentation with high magnification.

Androsace elongata L. (Behrendsen s.n.)
6. Outline in equatorial view.

Androsace maxima L. (Keinitzen 4197)
7. Ornamentation.
8. Outline in polar view.
9. Outline in equatorial view.
10. Rectangular endoaperture.

Cyclamen purpurascens Miller (Biol. exc. 1963—4315)
11. Outline in equatorial view.
12. Pole with small apocolpium.
13. Outline in equatorial view.
14. Rectangular endoaperture.
15. Outline in polar view; pole with large apocolpium.

PLATE IV (p.59)

Cyclamen hederifolium Aiton (Stefani s.n.)
1. Colpus and endoaperture.
2,3. Outline in equatorial view.
4. Outline in polar view.

Glaux maritima L. (Roosma s.n., except fig.8: Biol. exc. 1970—196)
5. Outline in equatorial view.
6. Heavy costae around slit-like ectocolpus and rather broad endoaperture; endoaperture with incised margins.
7. Outline in polar view; thick nexine.
8. Ornamentation in mesocolpium.
9. Colpus ends.
10. Endoaperture with horns.

Hottonia palustris L. (fig.11: France s.c., s.n.; fig.12: Leeuwenberg 237)
11. Long-styled specimen, narrow ectocolpus with one endoaperture.
12. Short-styled specimen, ectocolpus with two endoapertures.

PLATE V (p.60)

Hottonia palustris L. (Leeuwenberg 237) (short-styled specimen)
1. Outline in polar view.
2. Colpus.
3. Reticulum.

Hottonia palustris L. (France s.c., s.n.) (long-styled specimen)
4. Outline in polar view.
5. Outline in equatorial view.
6. Reticulum.

Primula stricta Horn. (Wikström s.n.)
7. Anastomosing colpi with apocolpial field.
15. Ornamentation.
16. Irregular outline.

Primula farinosa L. (Waid s.n.)

8. Outline in polar view of short-styled flower.
9. Outline in equatorial view of short-styled flower.
10. Outline in equatorial view of long-styled flower.
11. Outline in polar view of long-styled flower.
12. Apocolpial field of long-styled flower.
13. Ectocolpus.
14. Apocolpial field, short-styled flower.

PLATE VI (p.*61*)

Lysimachia vulgaris L. (Dieleman 384)
1. Outline in equatorial view, without fastigium.
3. Apocolpium at low focus.
4. Apocolpium at high focus.
5. Reticulum in mesocolpium.
6. Ectocolpus, tapering endoaperture and margo.
Lysimachia nummularia L. (Hessel, Klein and Rubers 1364)
2. Outline in equatorial view.
7. Reticulum in mesocolpium.
8. Reticulum at low focus.

PLATE VII (p.*62*)

Lysimachia vulgaris L. (Dieleman 384)
1. Outline in polar view.
Lysimachia terrestris Britton, Sterns and Pogg.(Mennega s.n.)
2. Tapering endoaperture, distinct margo.
3. Outline in polar view.
4. Reticulum in mesocolpium.
5. Outline in equatorial view.
9. Microreticulum in apocolpium.
Lysimachia punctata L. (Hinterbuber 459)
6. Outline in polar view.
7. Reticulum in apocolpium, distinct margo.
8. Reticulum at low focus.

PLATE VIII (p.*63*)

Lysimachia thyrsiflora L. (Leeuwenberg s.n.)
1. Outline in equatorial view with distinct fastigium.
2. Reticulum at low focus.
3. Reticulum at high focus.
4. Reticulum in apocolpium.
Lysimachia ephemerum L. (Biol. exc. 1967—2180)
5. Colpus, endoaperture with horns.
6. Outline in equatorial view.
7. Reticulum.
8. Anastomosing horns.
9. Microreticulum at apocolpium.
Lysimachia nemorum L. (Punt s.n.)
10. Outline in equatorial view.

PLATE IX (p.*64*)

Primula vulgaris Huds. (Schütz s.n.) short-styled flower

1. Outline in equatorial view.
2. Outline in polar view.
3. Rather distinct colpus at low focus.
4. Colpus at high focus.
5. Reticulum.

Primula vulgaris Huds. (Douw and Van Embden 30) long-styled flower
6. Outline in polar view.
7. Colpus and indistinct endocracks.

Lysimachia nemorum L. (Punt s.n.)
8. Tectum perforatum or microreticulum.
9. Slit-like colpus and tectum perforatum.
10. Endoaperture with distinct horns.
11. Outline in polar view.

PLATE X (p.65)

Primula elatior (L.) Hill (Hekking s.n.) short-styled flower
1. Outline in equatorial view.

Primula elatior (L.) Hill (H.F. Smith 2002) long-styled flower
2. Outline in polar view.
3. Outline in equatorial view.
7. Colpus and ornamentation.

Primula veris L. (Hessel, Klein and Rubers 114)
4. Indistinct colpus at low focus.
5. Indistinct colpus at high focus, reticulum.
6. Outline in polar view.

Primula hirsuta All. (Behrendsen s.n.)
8. Tectum perforatum.
9. Outline in polar view.
10. Colpus with indistinct endoaperture, without costae.
11. Outline in equatorial view.

PLATE XI (p.66)

Primula scotica Hook. (Barkman 2392)
1. Outline.
2. Anastomosed colpi.
3. Ectocolpus with indistinct, small endoaperture.

Samolus valerandi L. (fig.4: Oosterveld 267; figs.5—6: Teunissen and Wildschut s.n.)
4. Ectocolpus with indistinct, small endoaperture.
5. Outline in equatorial view, small fastigium.
6. Outline in polar view.

Steironema lanceolata (Walter) A. Gray (Lejeune 1010)
7. Outline in polar view.
8. Outline in equatorial view.
9. Outer ends of the colpi and tectum.
10. Ectocolpus.
11. Endoaperture, indistinct.

PLATE XII (p.67)

Trientalis europaea L. (figs.1—2: Nederland s.n., s.c.; figs.3—5: Biol. exc. 1970—1653)
1. Tectate to microreticulate ornamentation.
2. Outline in polar view.
3. Wart-like processes on the nexine and indistinct shallow colpi without distinct margins.

4. Ornamentation.
5a and b. Sections showing the processes on the nexine, long, slender columellae and slightly elongated capita.

PLATE XIII (p.*68*)

Soldanella alpina L. (Behrendsen s.n.)
1. Outline in equatorial view.
2. Outline in polar view.
3. Rather broad bridge.
7. Apocolpium at high focus.
8. Apocolpum at low focus.
Trientalis europaea L. (figs.4,6: Biol. exc. 1970—1653; fig.5: Nederland s.n., s.c.)
4. Ectocolpus.
5. Ectocolpus.
6. Outline of irregular pollen grain.
Samolus valerandi L. (Teunissen and Wildschut s.n.)
9. Ornamentation at low focus.
10. Ornamentation at high focus.

PLATE XIV (Scanning electron micrographs; p.*69*)

Glaux maritima L. (Roosma s.n.)
1. Equatorial view ($\times$ 3200).
2. Structure of mesocolpial ornamentation; tectum perforatum ($\times$ 5000).
Other species
3. *Primula elatior* (L.) Hill (Hekking s.n.). Equatorial view; narrow colpi ($\times$ 4000).
4. *Lysimachia nummularia* L. (Hessel, Klein and Rubers 1364). Structure ($\times$ 5000).
5. *Samolus valerandi* L. (Teunissen and Wildschut s.n.). Polar view and structure ($\times$ 5000).
6. *Hottonia palustris* L. (Leeuwenberg 237). Polar view and structure ($\times$ 4700).

PLATE I

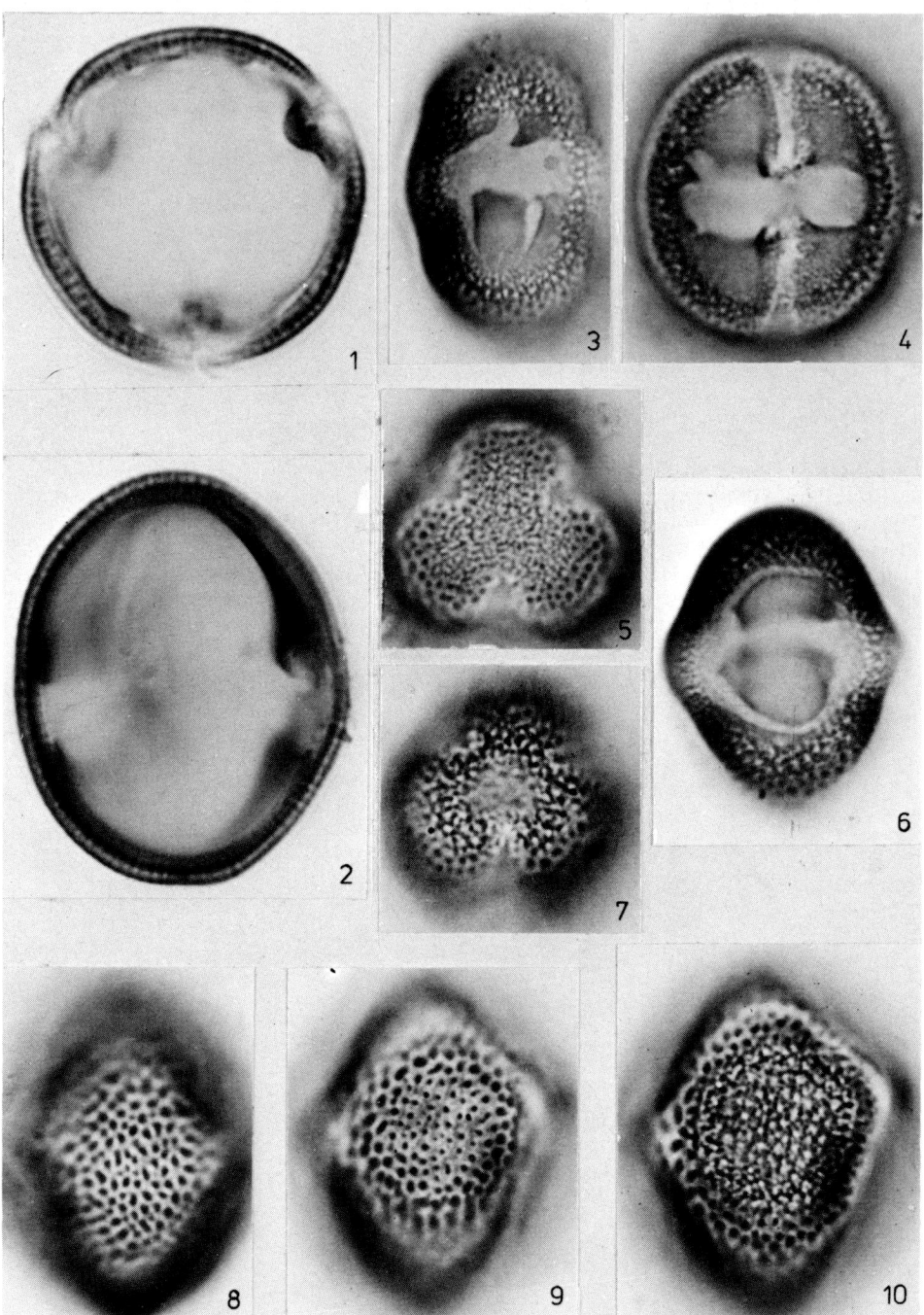

PLATE II

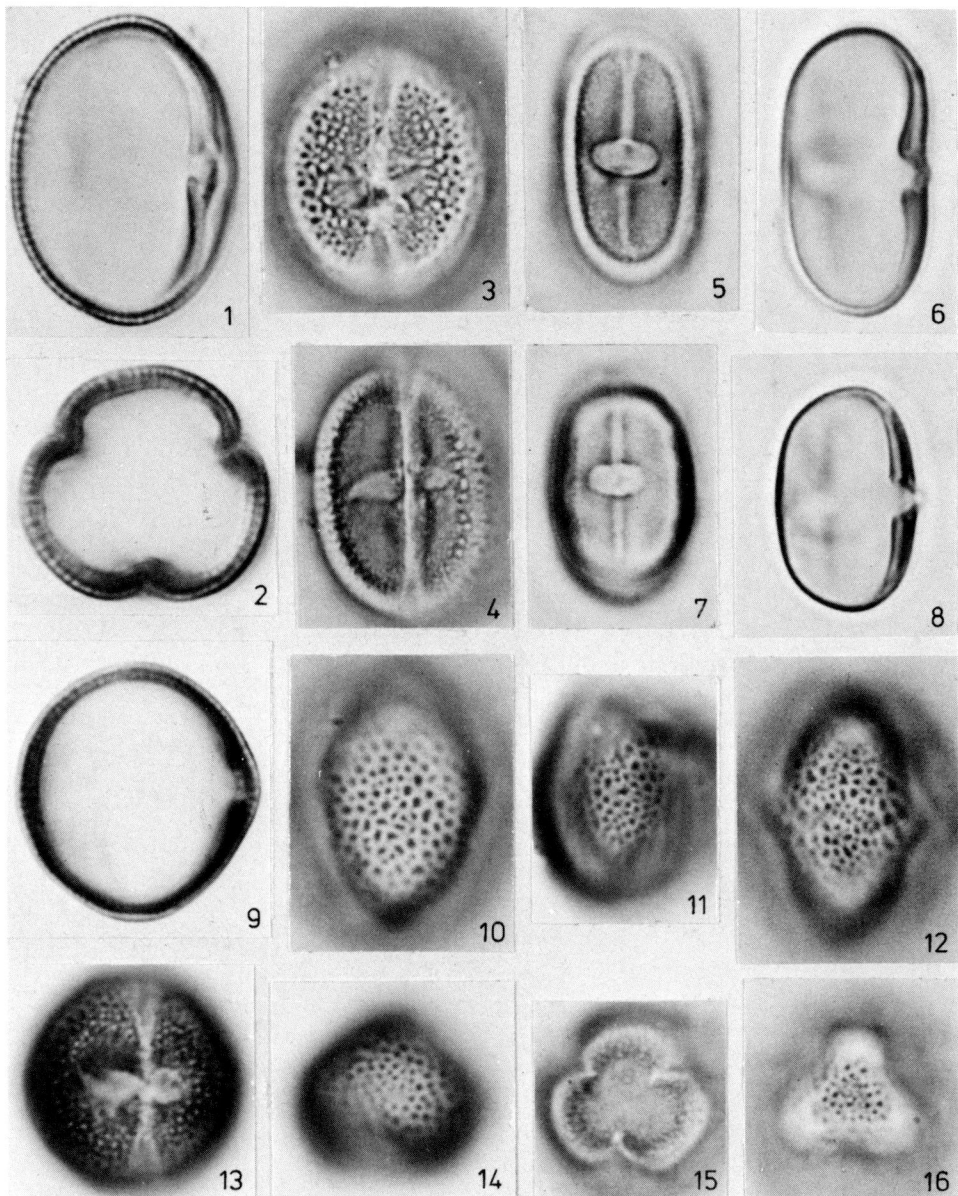

58

PLATE III

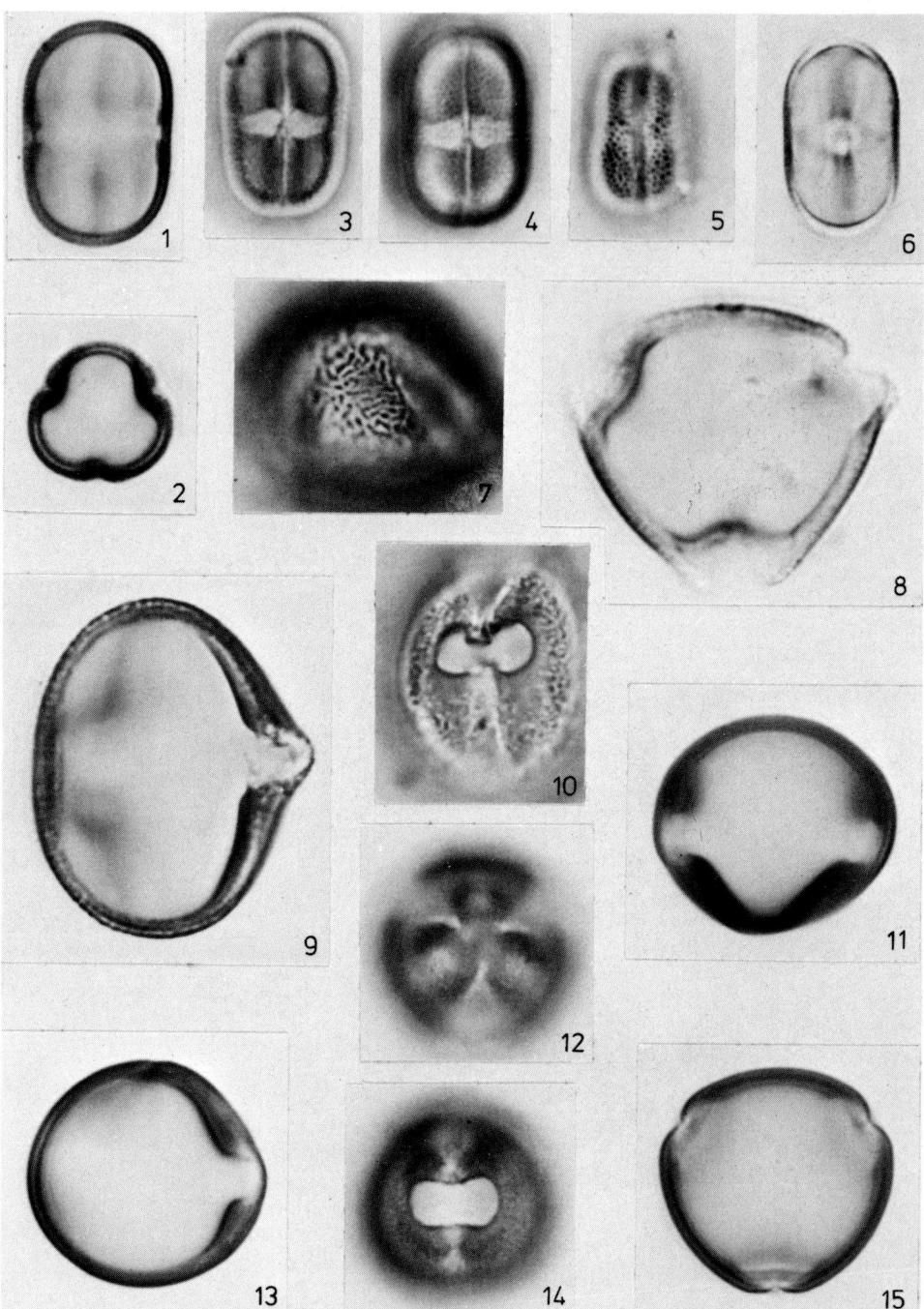

PLATE IV

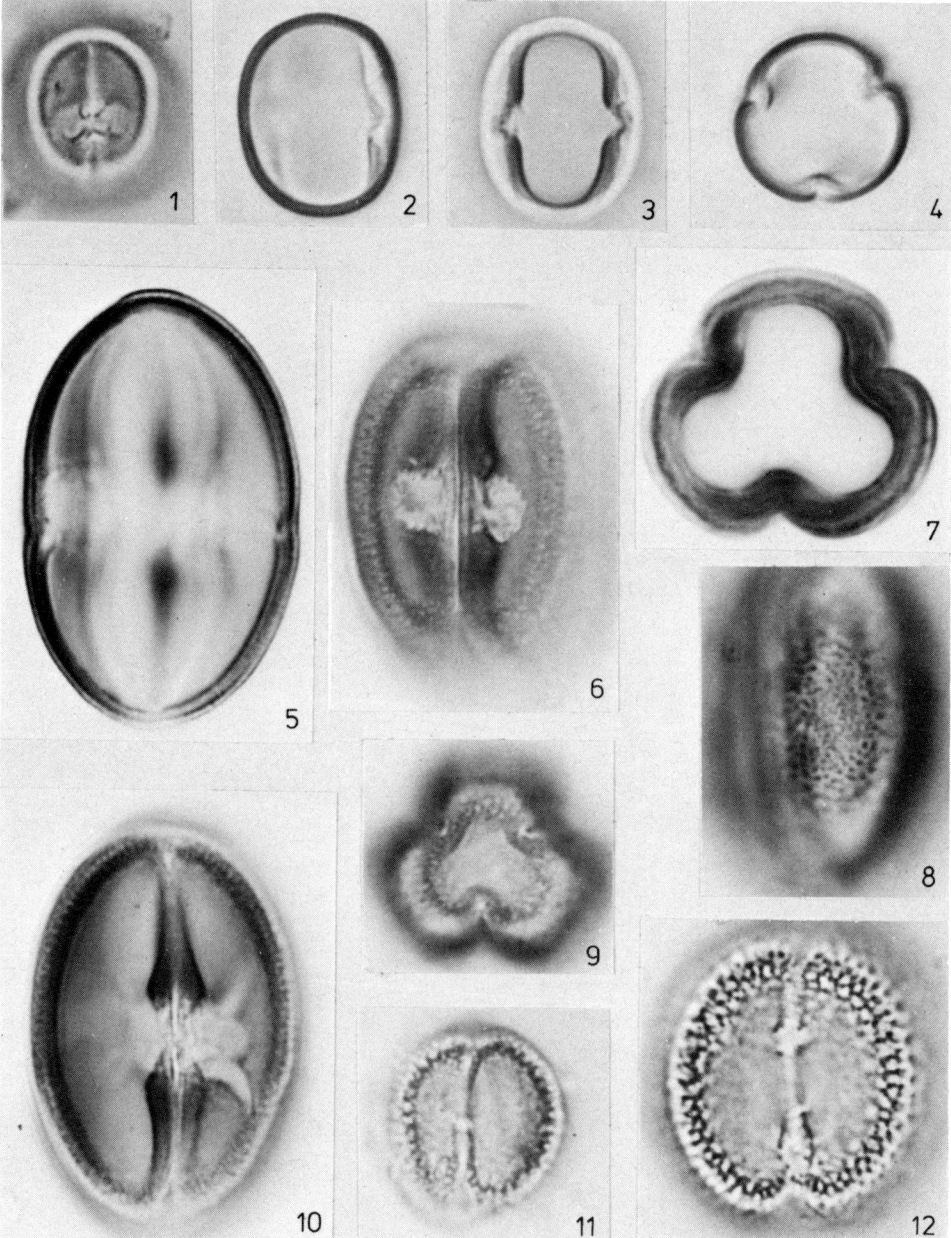

PLATE V

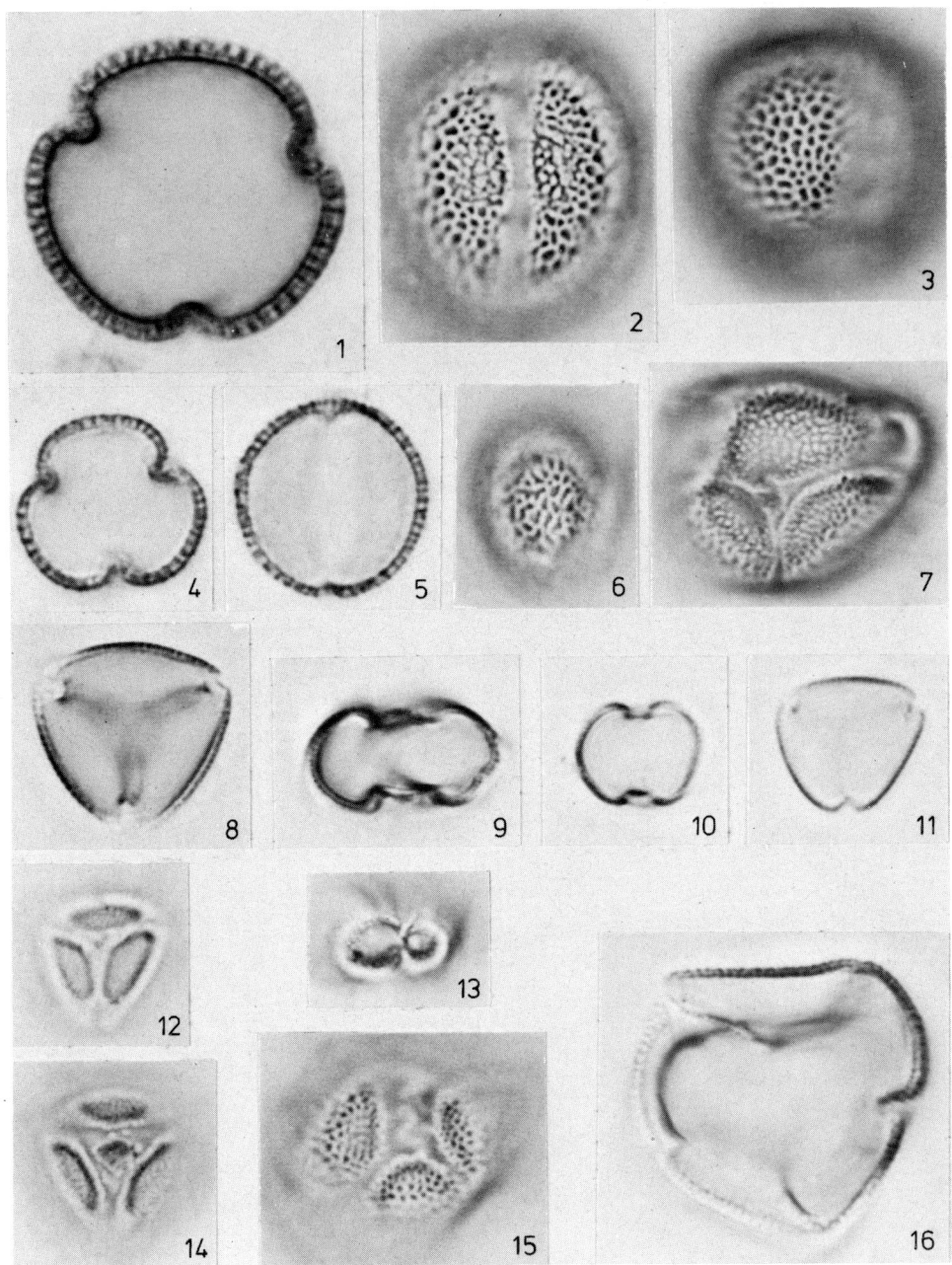

PLATE VI

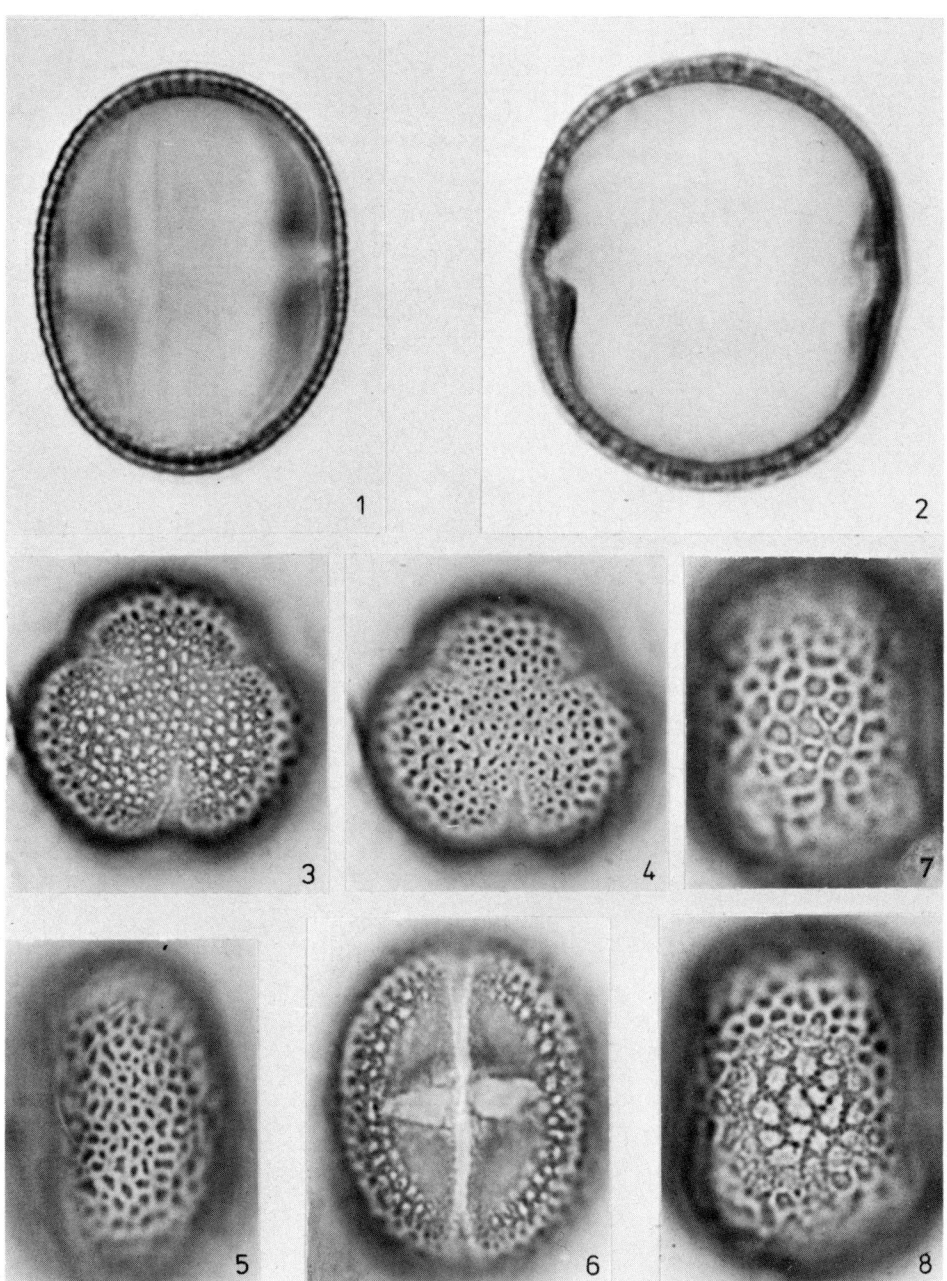

PLATE VII

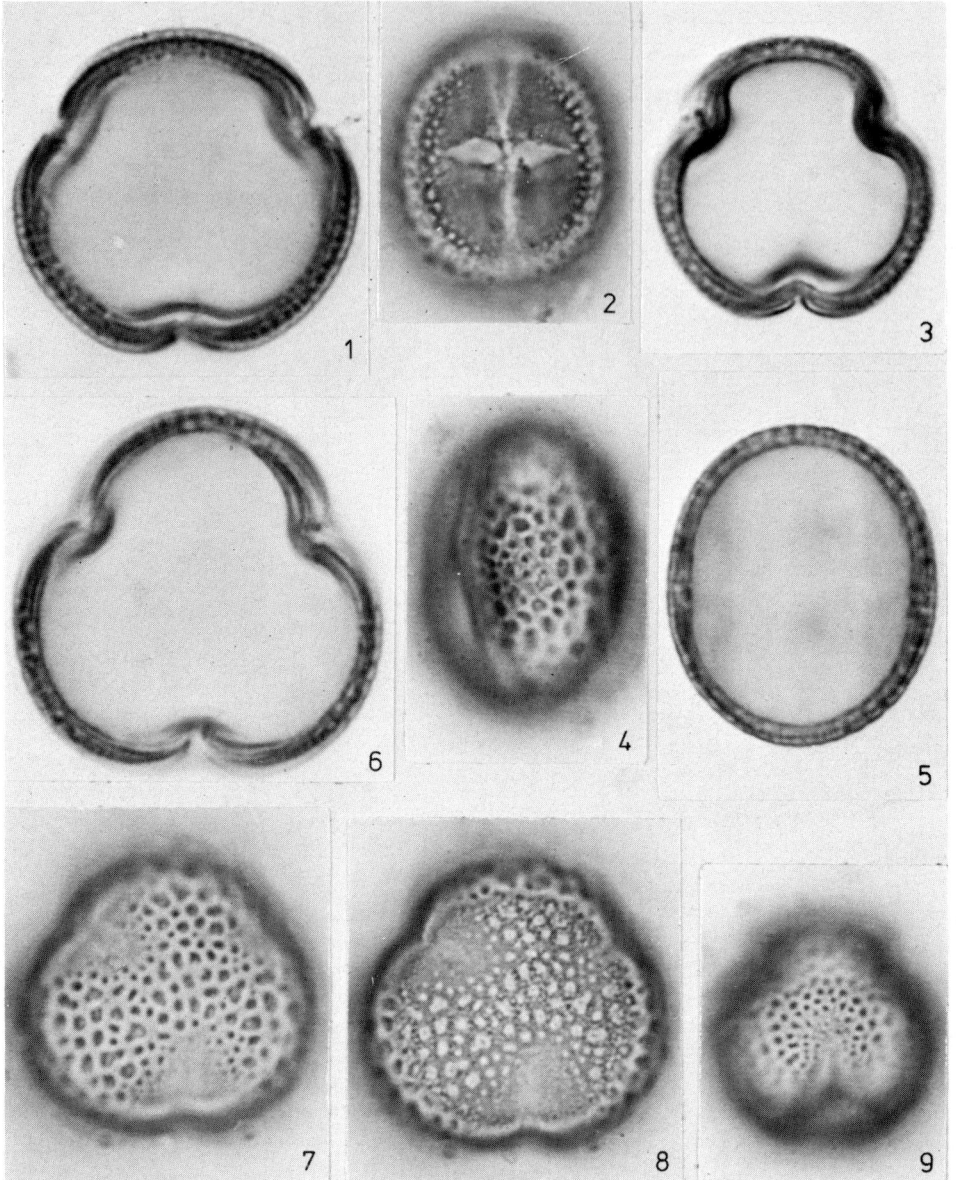

PLATE VIII

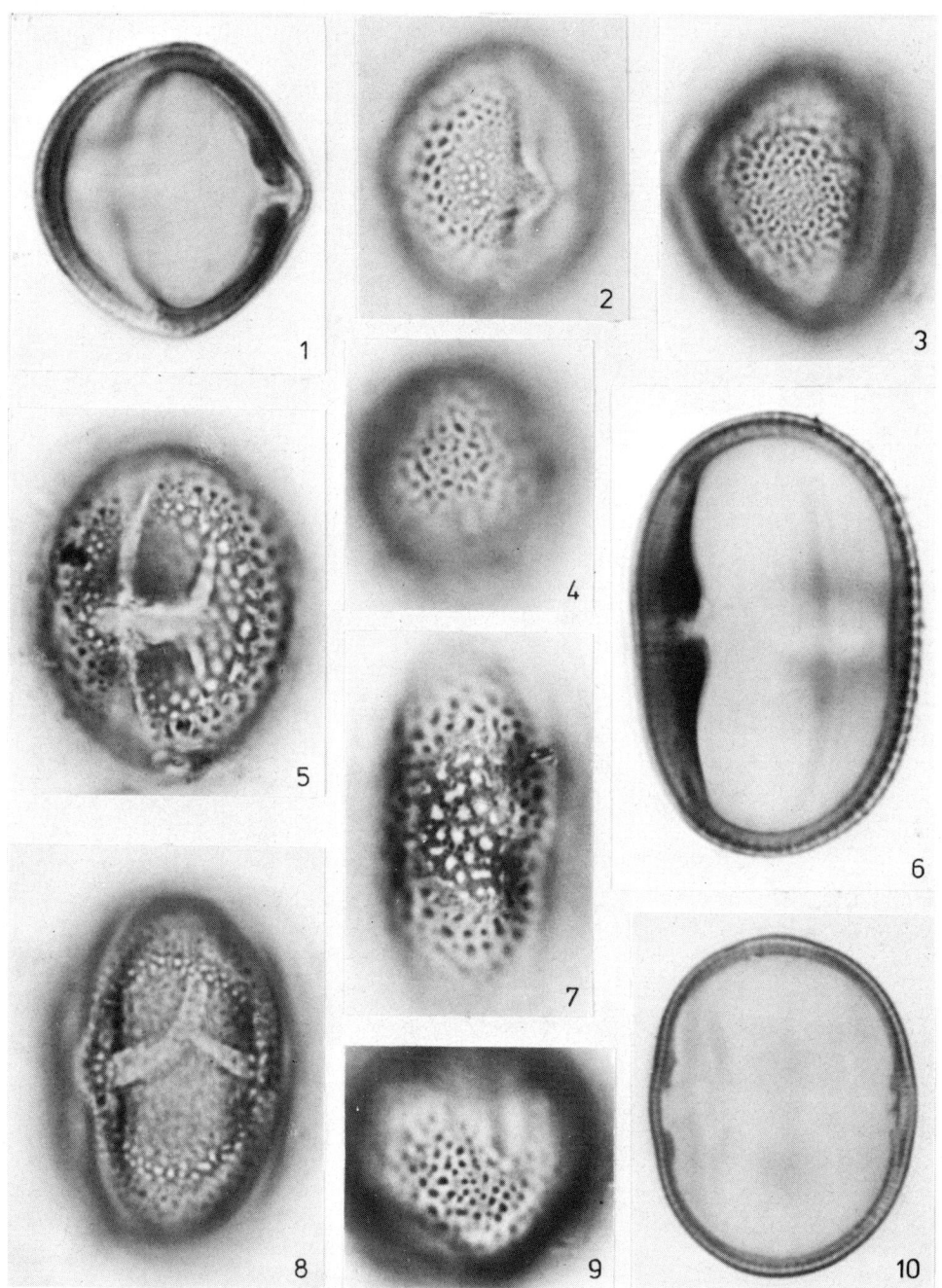

PLATE IX

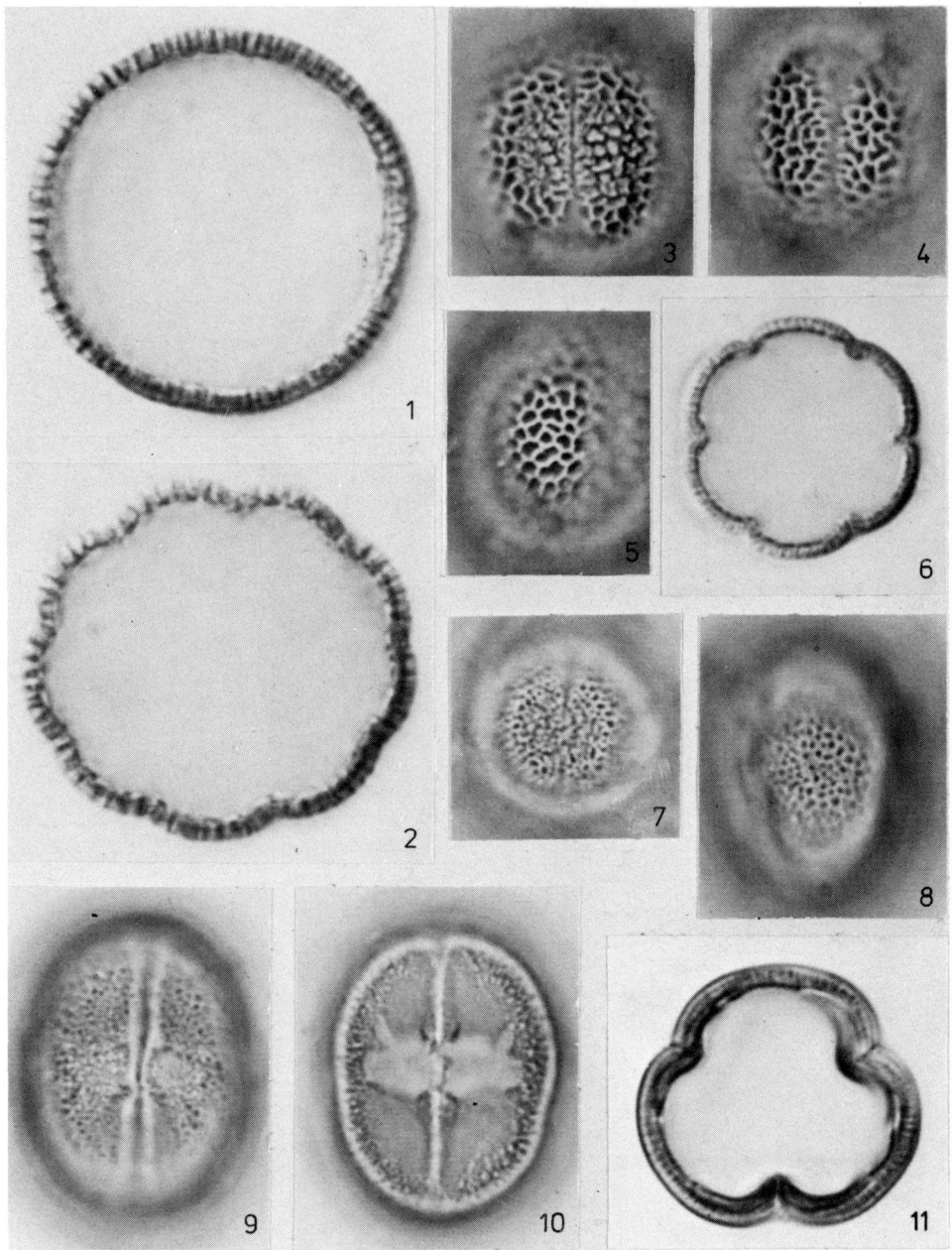

PLATE X

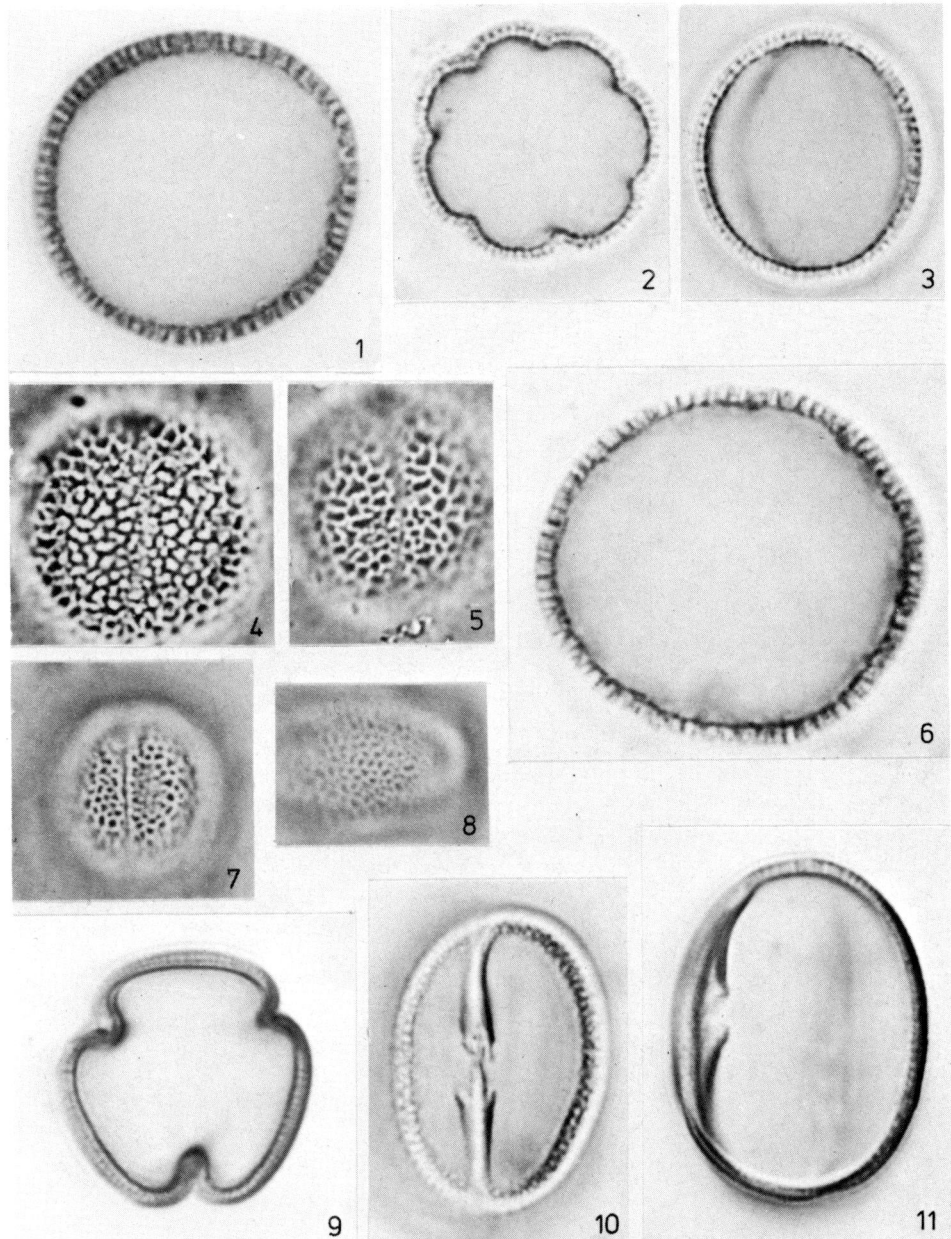

PLATE XI

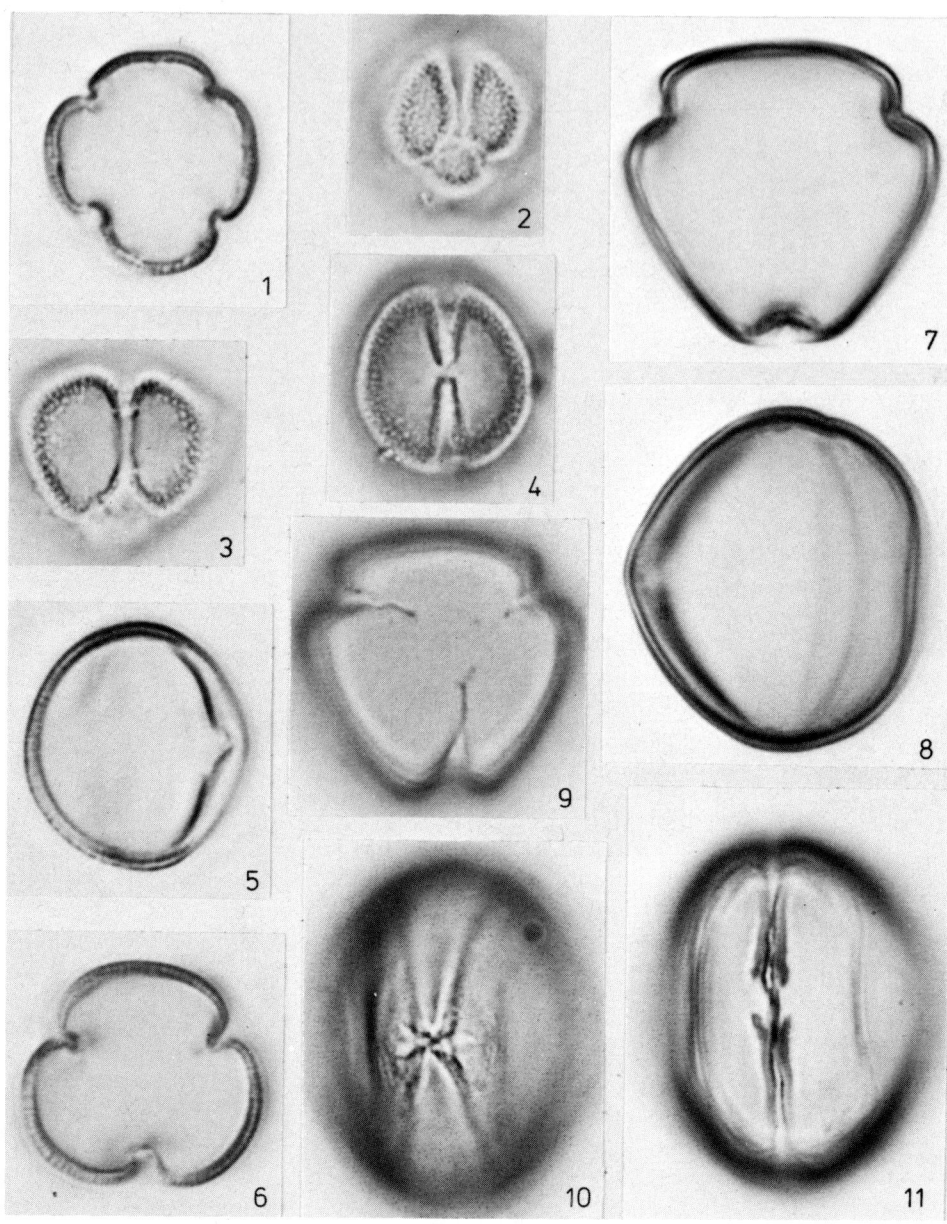

PLATE XII

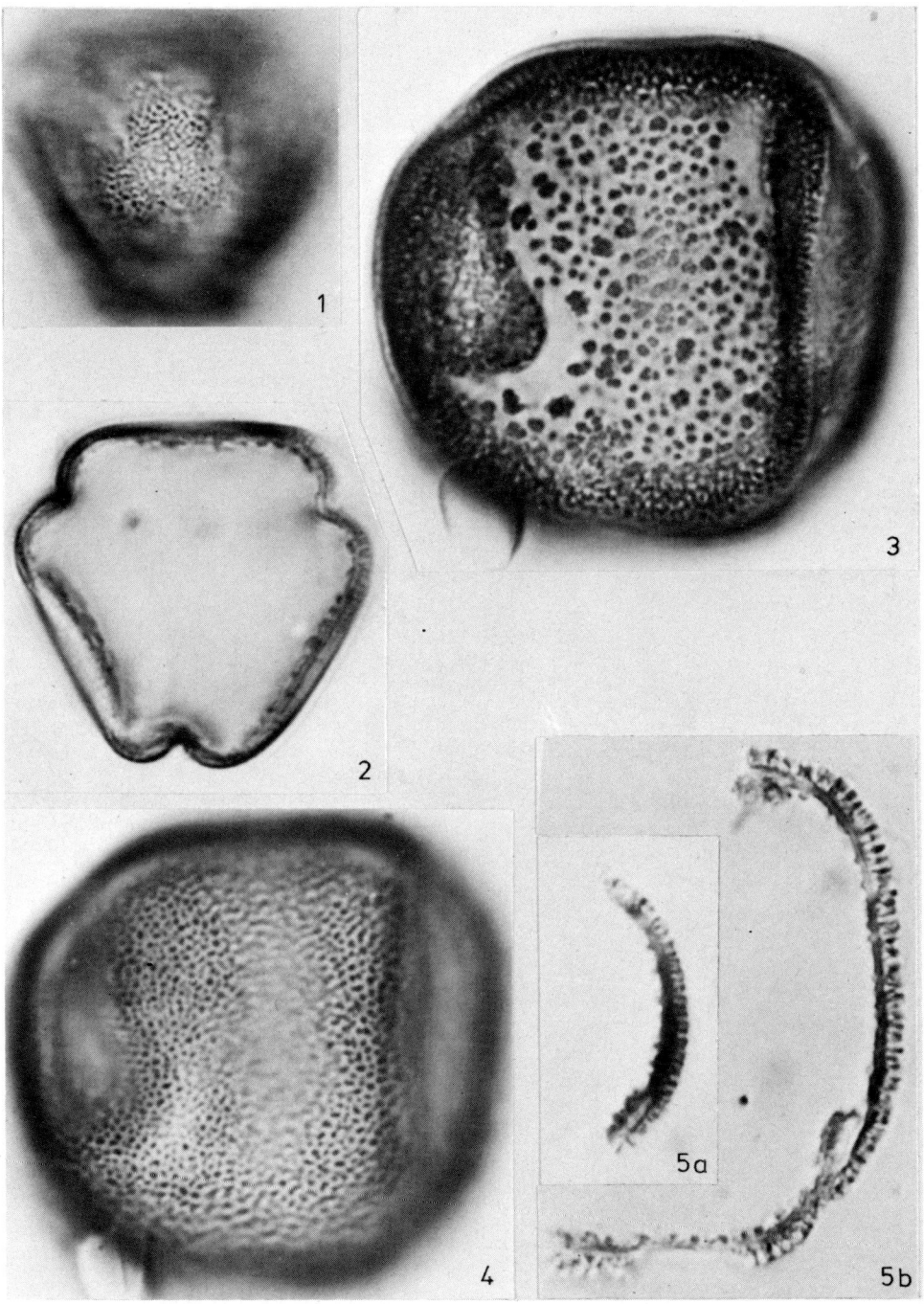

PLATE XIII

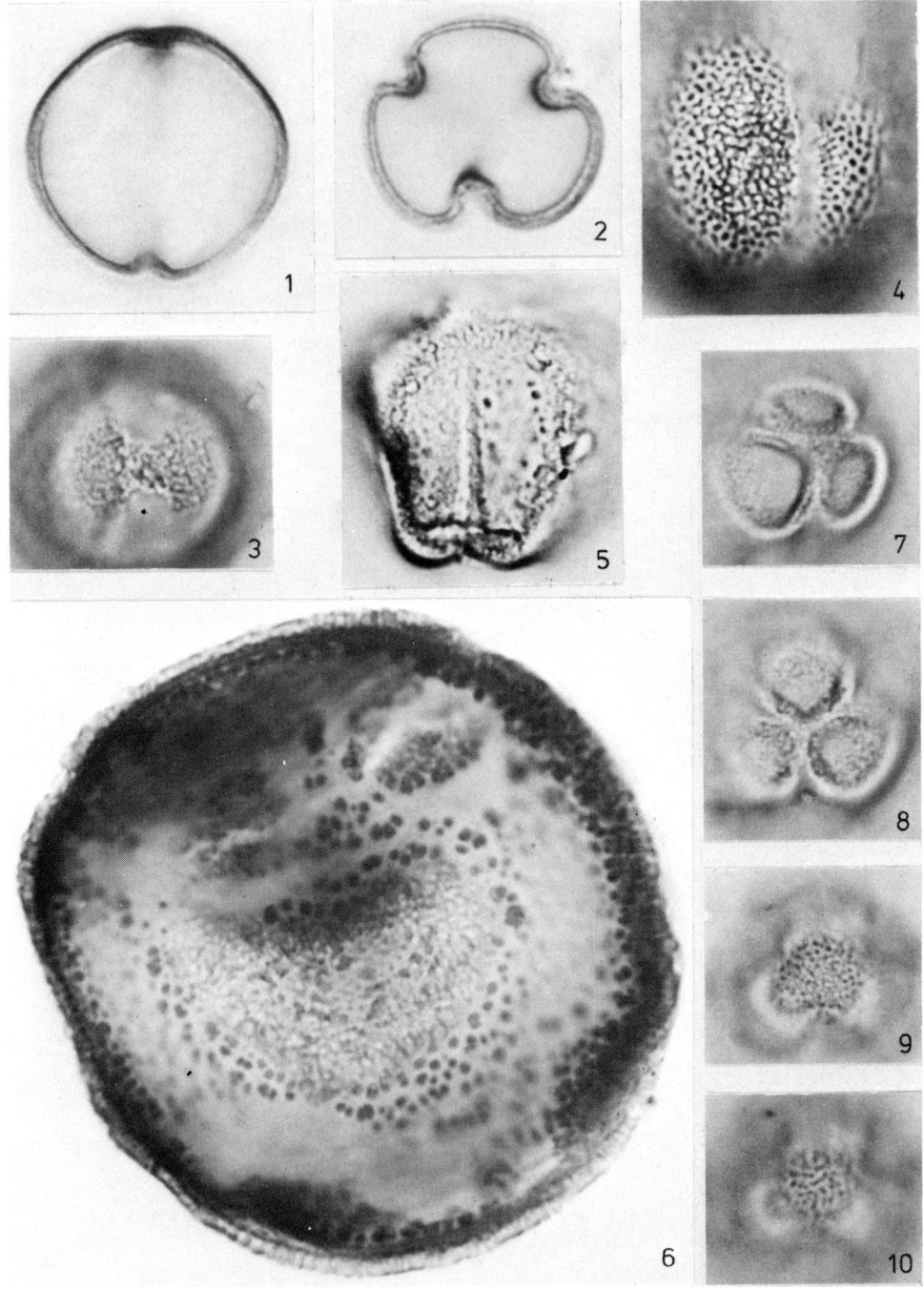

PLATE XIV

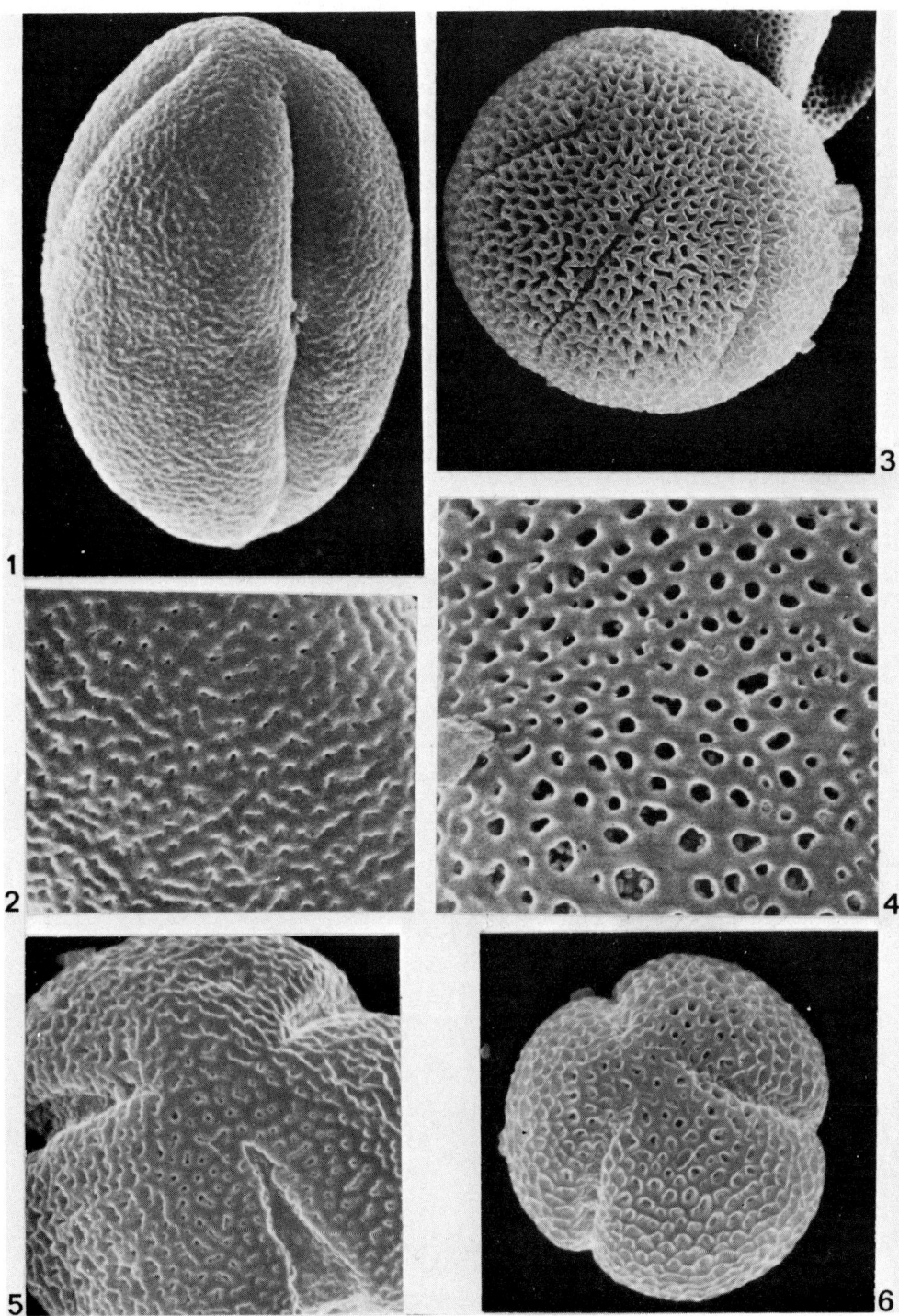

REFERENCES

Beug, H., 1961. Leitfaden der Pollenbestimmung, 1. Fischer, Stuttgart, 92 pp.
Erdtman, G., 1952. Pollen Morphology and Plant Taxonomy. Angiosperms. Almquist and Wiksell, Stockholm, 539 pp.
Erdtman, G., Berglund, B. and Praglowski, J., 1961. An introduction to a Scandinavian Pollen Flora, I. Almquist and Wiksell, Stockholm, 92 pp.
Faegri, K. and Iversen, J., 1950. Textbook of Modern Pollen Analysis. Munksgaard, Copenhagen, 169 pp.
Faegri, K. and Iversen, J., 1964. Textbook of Pollen Analysis. Munksgaard, Copenhagen, 237 pp.
Hiirsalmi, H., 1969. *Trientalis europaea* L. A study of the reproductive biology, ecology and variation in Finland. Ann. Bot. Fenn., 6: 119—140.
Huynh, Kim-Lang, 1970. Le pollen et la systématique chez le genre *Lysimachia* (Primulaceae). I. Morphologie générale du pollen et palynotaxonomie. Candollea, 25: 267—296.
Iversen, J. and Troels-Smith, J., 1950. Pollenmorphologische Definitionen und Typen. Dan. Geol. Unders., IV, 3(8): 1—53.
Oldfield, F., 1959. The pollen morphology of some of the West European Ericales. Pollen Spores, 1: 19—49.
Praglowski, J. and Punt, W., 1974. An elucidation of the microreticulate structure of the exine. Grana, 13: 45—50.
Reitsma, Tj., 1966. Pollen morphology of some European Rosaceae. Acta Bot. Neerl., 15: 290—307.
Spanowsky, W., 1962. Die Bedeutung der Pollenmorphologie für die Taxonomie der Primulaceae—Primuloideae. Feddes Report, 65: 149—215.
Tarnavaschi, I. T. and Mitroiu, H., 1960. Considérations palynologiques sur les représentants de la famille des Primulacées de la flore Roumaine. Comun. Acad. Rep. Pop. Rom., 10: 111—118.
Wendelbo, P., 1961. Studies in Primulaceae. III. On the genera related to *Primula* with special reference to their pollen morphology. Arb. Univ. Rergon. Nat.-Naturv. ser., 1961., 19: 1—31.
Yordanov, D. and Peev, D., 1970. An investigation of the three species *Primulas* from section Aleurita Duby distributed in Bulgaria. Mitt. Bot. Inst., 20: 131—150.

The Northwest European Pollen Flora, 4

ADOXACEAE

TJ. REITSMA[1] and ALBERDINA A. M. L. REUVERS

Laboratory of Palaeobotany and Palynology, State University, Utrecht (The Netherlands)

LITERATURE

Erdtman (1952, 1966), Erdtman et al. (1961, 1963), Faegri and Iversen (1950, 1964), Kuprianova and Alyoshina (1972).

SPECIMENS EXAMINED

Adoxa moschatellina L. — Belgium: Lejeune et Courtois 594 (U), Smit 2011 (U); Finland: Pankakoski s.n. (U); The Netherlands: Punt s.n. (U), Punt Anno 1959 (Fresh material) 2X; West Germany: Van den Burgh s.n. (U).

DESCRIPTION OF THE POLLEN TYPES

Adoxa moschatellina type (Plate I)

Pollen Class: (2-)3-Zonocolporate or syncolporate.
P/E Ratio: Suberect to semi-erect.
Apertures: Ectoaperture — colpus, long to very long, broad, deeply sunken; ends acute to slightly obtuse; bridge sometimes present; colpus membrane slightly granulate; fastigium present. Endoaperture — lalongate porus, indistinct; best visible in polar view; no costae present.
Exine: thin; sexine as thick as nexine; sexine decreasing in thickness towards the colpi and the apocolpium; sexine 1(columellae) thicker than sexine 2 (capita); columellae distinct; capita more or less depressed at the top.
Ornamentation: Reticulate to microreticulate; lumina decreasing in size towards the apocolpium and the ectocolpi, irregular in size and shape; muri simplicolumellate; columellae circular to slightly elliptic in outline.
Outlines: Equatorial view — elliptic. Polar view — 3-angular with convex sides and apertures situated in the obtuse angles.
Measurements: Glycerine jelly — P(longest axis) 22—26μ; E 19—22μ; P/E ratio 1.05—1.28; apocolpium index 0.10—0.25, sometimes 0; exine about 1.5μ; lumina about 1μ. Silicone oil — P(longest axis) 22—26μ; E 18—22μ; P/E ratio 1.07—1.28.
Species: Adoxa moschatellina L.

---

[1] Present address: Rijksdienst voor de Ysselmeerpolders, Lelystad (The Netherlands).

Comments

The pollen grains of this type can easily be confused with those occurring in the *Sambucus nigra* type (Punt et al., 1974). These two types have to be compared carefully. The most important difference, always recognizable at 400 × magnification, is the colpus membrane. A summary of the differences is given in Table I.

TABLE I

The differential characters of the *Adoxa moschatellina* type and the *Sambucus nigra* type

Adoxa moschatellina type	*Sambucus nigra* type
colpus membrane granulate	colpus membrane nudate
fastigium present	fastigium absent
lalongate endoporus (rather indistinct)	endocolpus (rather indistinct)
pollen (2-), 3-zonocolporate or syncolporate	pollen (2-), 3-zonocolporate, never syncolporate

REFERENCES

Erdtman, G., 1952. Pollen Morphology and Plant Taxonomy (An Introduction to Palynology, 1. Angiosperms). Almqvist and Wiksell, Stockholm, 539 pp.
Erdtman, G., 1966. Pollen Morphology and Plant Taxonomy — Angiosperms. Hafner, New York, N.Y., 2nd ed., 553 pp.
Erdtman, G., Berglund, B. and Praglowski, J., 1961. An Introduction to a Scandinavian Pollen Flora, I. Almqvist and Wiksell, Stockholm, 92 pp.
Erdtman, G., Praglowski, J. and Nilsson, S., 1963. An Introduction to a Scandinavian Pollen Flora, II. Almqvist and Wiksell, Stockholm, 89 pp.
Faegri, K. and Iversen, J., 1950. Textbook of Modern Pollen Analysis. Munksgaard, Copenhagen, 169 pp.
Faegri, K. and Iversen, J., 1964. Textbook of Pollen Analysis. Munksgaard, Copenhagen, 2nd ed., 237 pp.
Kuprianova, L. A. and Alyoshina, L. A., 1972. Pollen and Spores of Plants from the European Part of the USSR. I. Izd. Akad. Nauk S.S.S.R., Leningrad, 171 pp (in Russian).
Punt, W., Reitsma, Tj. and Reuvers, A. A. M. L., 1974. Northwest European Pollen Flora. Caprifoliaceae. Rev. Palaeobot. Palynol., 18(1974):NEPF

PLATE I (× 1500, except fig.11)

Adoxa moschatellina L. (fig.1,2 and 11: Pankakosi s.n.; all other figs.: Smith 2011)
 1. SEM-photograph, equatorial view with domed sexine in equatorial plane.
 2. SEM-photograph, equatorial view, aberrant shape which often occur.
 3. Equatorial view, cross-section.
 4. Polar view, cross-section.
 5. Polar view, apocolpium, syncolpate grain, lumina distinctly decreasing towards the pole.
 6. Fastigium at high focus.
 7. Fastigium at low focus.
 8. Granulate colpus membrane.
 9. Reticulum in mesocolpium at high focus, large lumina intermixed with smaller ones.
10. Reticulum in mesocolpium at low focus.
11. SEM-photograph, reticulum in mesocolpium (× 3750).

PLATE I

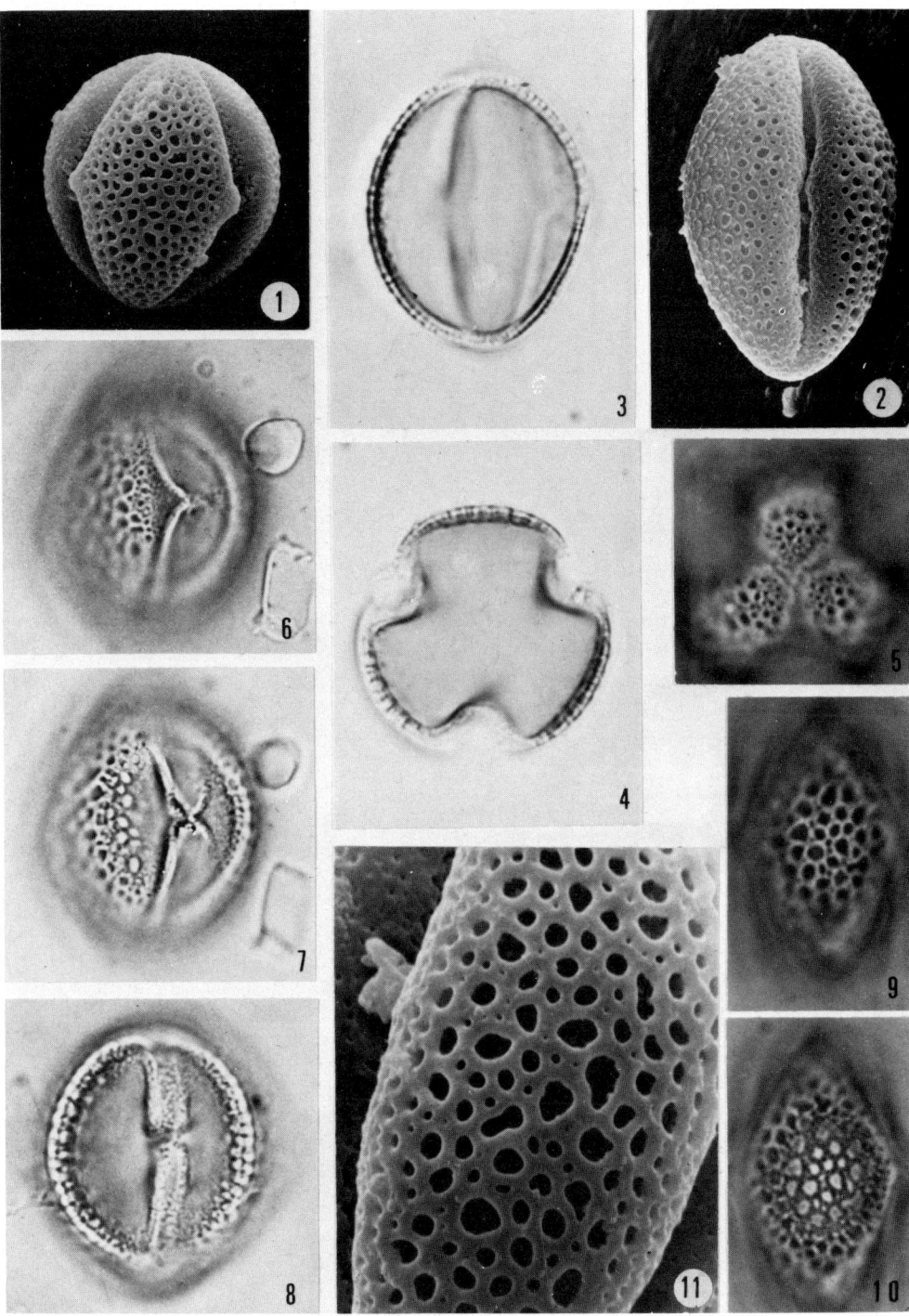

The Northwest European Pollen Flora, 5

SPARGANIACEAE AND TYPHACEAE

W. PUNT

Laboratory of Palaeobotany and Palynology, State University, Utrecht (The Netherlands)

LITERATURE

Beug (1961), Erdtman (1943, 1952), Erdtman et al. (1961, 1963), Faegri and Iversen (1950, 1964), Visset (1971), Wodehouse (1932, 1935).

INTRODUCTION

Unlike previous parts of the Northwest European Pollen Flora the present part deals with two families together. The results show, that the pollen grains of *Typha angustifolia* are so similar to those of some *Sparganium* species that they have been included in a single pollen type. Also the two genera are taxonomically so close to one another that it seems not only logical but merely practical to combine them in one part.

SPECIMENS EXAMINED

Sparganium angustifolium Michaux — Finland: Vervoort 268(U); France: Billot s.n.(L), Bouchard s.n.(L); Portugal: S. Fernandes et al. 4491(U), Rainha et al. 2175(U); The Netherlands: Docters Van Leeuwen 6796(U), Jansen en Wachter 18.478/79(L), Van der Voo 5964a(L), Van der Voo s.n.(L).
Sparganium emersum Rehman — Finland: Marklund s.n.(U); Great Britain: Howard 756(U); The Netherlands: Mennega s.n.(U), Van Steenis s.n.(U), Vegter s.n.(U), s.c, s.n. Botanical Museum 88587A(U); West-Germany: Stafleu s.n.(U).
Sparganium erectum L. ssp. *erectum* — Great Britain: Howard 757(U); The Netherlands: Buysman 1647(U), Gottenbos s.n.(U), Lieftinck s.n.(U), Punt (fresh material) 1964.
Sparganium erectum L. ssp. *microcarpum* (Neumann ex Krok) Hylander — The Netherlands: Brand s.n.(L), Hunger s.n.(L), Oudemans 1203(L), Van der Voo 57.060(L).
Sparganium erectum L. ssp. *neglectum* (Beeby) Schinz et Thellung — Austria: Porta s.n. (Herb. Buysman 812) (U); Sweden: Samuelsson 67(L); The Netherlands: Jansen en Wachter 18480/81(L), Kern en Reichgelt 4845(L), A. G. de Wilde 56(L).
Sparganium fluitans (Morong) Robinson — Canada: Rolland-Germain 2868(L).
Sparganium friesii Beurling — Sweden: Weimarck s.n.(S), s.c., s.n. Anno 1922(S); U.S.S.R.: Fischer s.n.(L), Karo 488(L).
Sparganium glomeratum Laestiboudois — Norway: Biol. exc. 1965—245(U); Sweden: Brotheriez s.n.(L), Runne s.n.(S).
Sparganium minimum Wallroth — Germany: Gerhardt s.n.(L); Great Britain: Kirk 1144(L); The Netherlands: Hekking et al. s.n.(U), Holkema s.n.(L), Lanjouw s.n.(U), Van de Sande et al. s.n.(L), Van Steenis s.n.(U), Swart 790(U). Voorrips (fresh material) 1964.

Typha angustifolia L. — Sweden: Asplund s.n.(S), H. Fries s.n.(S); The Netherlands: Bakhuizen Van de Brink 4964(U), Van Dijk s.n.(U), Van Ooststroom 24690(U), Pfaeltzer s.n.(U), Van Royen 729(U), Rem (fresh material) 1951.
Typha latifolia L. — Austria: Porta s.n. (Herb. Buysman 818) (U); France: s.c., s.n. Montpellier 20224G; Germany: Behrendsen 2861(U); The Netherlands: Van Dijk s.n.(U), Rem (fresh material) 1951, Voorrips (fresh material) 1964.
Typha minima Funk — France: Lardière s.n.(L); Germany: Lahn s.n.(U); Austria: Behrendsen s.n.(U), Drobny s.n.(L).
Typha shuttleworthii Koch et Sonder — Rumania: Topa 2988(L); Switzerland: Shuttleworth s.n.(L).

KEY TO THE POLLEN TYPES

(N.B.: All features of the ornamentation mentioned in the keys are based on the structures in the area opposite to the aperture.)

1.a. Pollen grains in tetrads . . . . *Typha latifolia* type
   b. Pollen grains in monads 2
2.a. Tectate or distinctly microreticulate; lumina and perforations usually smaller than 1 µm, sometimes up to 1 µm . . . *Sparganium erectum* type
   b. Reticulate; lumina wider than 1 µm, often up to 3 µm or more . . . . *Sparganium emersum* type

DESCRIPTION OF THE POLLEN TYPES

Sparganium emersum type (Plate I, 6; Plate II; Plate III; Plate IV, 3—5)

Pollen class: Monoporate.
Apertures: Ectoaperture — porus, more or less circular in outline, slightly sunken; margins distinct or indistinct; porus membrane beset with sexine elements. Endoaperture — porus, outline following that of the ectoaperture and therefore more or less congruent with it; no costae.
Exine: Exine of normal thickness. Sexine about as thick as nexine, sometimes slightly thinner and less frequently slightly thicker. Sexine 1 usually thinner than sexine 2. Sexine 1 consists of low, thin columellae. Sexine 2 is semitectate.
Ornamentation: Reticulate. Muri thick or thin; in a few species the muri are parallel sided, but in some other species they are slightly thicker towards the base, simpli- or duplicolumellate, often interrupted; in L—O analysis the columellae are not always distinct, outline circular; in high magnification the muri may have small excrescences (in *Sparganium*) or may be smooth (*Typha angustifolia*). The lumina are fine or coarse, and they vary considerably in width and outline. The largest lumina are found in the area opposite to the porus area, and they decrease distinctly towards the porus area.
Outline: The overall shape is rather irregular and the outline depends on the position of the grain. Usually the grains are mainly elliptic in outline with one side more convex than the other, or even slightly angular. If the position of the pollen grain does not show this angular-elliptic outline, it may be circular. A complete range of outlines between the two extremes is possible.

Measurements: Glycerine jelly — longest axis 25—36 μm; exine between 1.5 and 2.5 μm; diameter of pori up to 3 μm. Silicone oil — longest axis 23—31 μm.
Species: Sparganium angustifolium, S. emersum, S. friesii, S. fluitans, S. glomeratum, S. hyperboreum, S. minimum, some specimens of *S. erectum* s.l. (Van der Voo 57.060, Jansen en Wachter 18480/81, Porta s.n.) and *Typha angustifolia*

Key to groups and species

1.a. Lumina large, largest ones usually ca. 2 μm or more in diameter; in L—O analysis columellae usually distinct 2
  b. Lumina rather small, usually 1—2 μm; in L—O analysis columellae usually distinct 3
2.a. Muri usually thin, usually not or only slightly interrupted; lumina often more or less elliptic in outline .... *Sparganium emersum*
  b. Muri usually thick, more or less interrupted; lumina irregular in outline .... *S. minimum* group
        (*Sparganium glomeratum, S. hyperboreum, S. minimum*)
3.a. Muri in short, irregular curved sections, forming a rugulate pattern, not distinctly reticulate; sexine pattern running into porus .... *Typha angustifolia*
  b. Ornamentation more or less distinctly reticulate; porus without distinct sexine pattern 4
4.a. Muri thin, usually simplicolumellate; lumina very angular in outline, usually angular-elongated .... *S. friesii* group
        (*Sparganium angustifolium, S. friesii, S. fluitans*)
  b. Muri thick, usually duplicolumellate; lumina more or less regular in outline, elliptic or circular .... some specimens of *S. erectum*

Comments

Wodehouse (1932) noted that the pollen grains of *Sparganium* species are virtually indistinguishable from those of *Typha angustifolia*. Beug (1961) and Faegri and Iversen (1964) do not give any critical differences between *Typha angustifolia* and several European *Sparganium* spp. Erdtman et al. (1961) also stated that pollen grains in the Sparganiaceae are of the same type as the monads in the Typhaceae.

In the present study most of the *Sparganium* species are combined with *Typha angustifolia* in one type. At a magnification of 400 × it was difficult to separate the pollen grains of these two genera, but at a higher magnification it was quite possible to differentiate them. *Typha angustifolia* is characterized by its rugulate pattern and also at very high magnification (SEM) the muri of the reticulum are smooth, whereas most *Sparganium* species show blunt excrescences. The SEM-photomicrograph which Visset (1971) has given of *Typha angustifolia* does not show the rugulate pattern, but the muri are smooth and without excrescences.

Apart from this differentiation of *Typha angustifolia* from other species in this pollen type, it was possible to separate the *Sparganium* species into two groups and two species. The differential characters are based on the width and outline of the lumina, the breadth of the muri and the presence of interrupted muri.

The *Minimum* group is best characterized by its thick and usually distinctly interrupted muri and, in addition to these two obvious characters, the pollen grains usually have indistinct pori.

The *Friesii* group, on the other hand, is best characterized by its rather small, irregular, often more or less elongated and angular lumina; the pori in this group are usually more distinct than those in the *Minimum* group.

The pollen grains of *Sparganium emersum* resemble those of the *Minimum* group very closely, but the muri are distinctly thinner and the lumina are often more regular.

The width of the lumina is always measured on the side of the grain opposite to the porus area. The width of the lumina around the pori and at the top of the ellipsoid may be considerably smaller.

Finally, the pollen grains of some specimens of *S. erectum* belonging to ssp. *microcarpum* and ssp. *neglectum* are differentiated by their more or less regular and rather small lumina together with the thick muri which are usually dupli-columellate. This group of pollen grains is quite different from those in the *Sparganium erectum* type and there are no transitional forms. This anomaly, which is partly taxonomic, cannot be solved within the scope of this project.

Sparganium erectum type (Plate I, 1—5)

Pollen class: Monoporate
Apertures: Ectoaperture — porus, more or less circular in outline, slightly sunken; margins rather distinct; porus membrane beset with granulate sexine elements. Endoaperture — porus, outline following that of the ectoaperture and therefore more or less congruent with it.
Exine: Exine varying in thickness, gradually thickening towards the porus area. Nexine about as thick as sexine, usually thicker near the porus. Sexine 1 thinner than sexine 2. Sexine 1 consists of short, thin columellae. Sexine 2 is a tectum or semitectum.
Ornamentation: Tectum perforatum or microreticulate. Muri thick, usually thicker than width of lumina; sometimes lumina wider than muri; muri thin at top and distinctly thicker towards the base. In L—O analysis columellae circular in outline; muri simpli- or duplicolumellate, at high magnification (SEM) provided with small and low excrescences. Perforations and lumina irregular in outline.
Outline: The overall shape is rather irregular and the outline depends on the position of the grain. Usually the pollen grains are mainly elliptic in outline with one side more convex than the other, or even slightly angular. If the position of the pollen grain does not show this angular-elliptic outline, it may be circular. A complete range of outlines between these extremes is possible.
Measurements: Glycerine jelly — longest axis 22—36 μm; exine ca. 1.5 μm;

diameter of the pori ca. 3 µm. Silicone oil — longest axis 21—26 µm.
Species: Sparganium erectum ssp. *erectum, S. erectum* ssp. *microcarpum, S. erectum* ssp. *neglectum.*

Comments

The *Sparganium erectum* type has not previously been recognized as distinct from the other species of *Sparganium*. In fact the type is characterized by a single differential character; viz. the tectum or microreticulum. This character, however, is so clear that the differentiation of the two types is justified. After a close examination of all slides this character appeared to be clear and constant and even visible at a magnification of 400 ×.

Some specimens of *Sparganium erectum* identified as ssp. *microcarpum* and ssp. *neglectum* (Van der Voo 57.060, Jansen en Wachter 18480/81, Porta s.n.) show lumina wider than 1 µm. These latter specimens have been placed in the *S. emersum* type (p.78).

The SEM-micrographs of Visset (1971) do not differ in size, shape or ornamentation features from those given here.

Typha latifolia type (Plate V, 1—2; Plate VI)

Pollen class: Tetrads. Single grains monoporate.
Apertures: Ectoaperture — porus, more or less circular in outline, slightly sunken; margins not sharply delimited; porus membrane beset with sexine elements similar to those occurring near the porus area, the ornamentation is running into the porus. Endoaperture — porus, outline following that of the ectoaperture and therefore more or less congruent with it.
Exine: Exine varying in thickness, gradually thicker towards the porus area. Nexine about as thick as sexine, sometimes slightly thicker, especially towards the porus area, often thinner than sexine in the area opposite to the porus area. Sexine 1 thinner than or about as thick as sexine 2. Sexine 1 composed of short columellae which are rather distinct in optical section. Sexine 2 is semitectate, consisting of capita which are more or less circular in outline.
Ornamentation: Reticulate, microreticulate or rugulate. Muri often interrupted, in *Typha minima* arranged in short, curved sections, simpli- or duplicolumellate; at high magnification without excrescences; in L—O analysis columellae distinct (*T. minima*) or indistinct. Lumina irregular in outline, often angular elongated. The single grains are attached to each other by several small connections. These connections are bridge-like, narrow and irregular in length and shape.
Outline: Single grains usually arranged in a more or less flat plane, often in squares or rectangles, but sometimes in a linear sequence. As the single grains are not exactly circular in outline but have a long and a short axis, the outline is more often rectangular with also longer and shorter sides. In the middle of the tetrad the fissurae are distinctly sunken; the angles are obtuse.
Measurements: Glycerine jelly — longest axis 30—55 µm; exine ca. 2 µm, up to 3 µm near the porus; porus ca. 5 µm wide. Silicone oil — 44—55 µm.

Species: *Typha latifolia, T. minima, T. shuttleworthii*

Key to the species

1.a. Ornamentation distinctly rugulate; muri distinctly interrupted in very short curved sections; in L—O analysis columellae rather distinct, circular in outline; nexine about as thick as sexine or often thicker .... *Typha minima*
  b. Ornamentation microreticulate or only slightly rugulate; in L—O analysis columellae usually distinct; nexine usually thinner than sexine except in the porus area .... *Typha latifolia*
          Typha shuttleworthii

Comments

Typha laxmannii Lepechine, mentioned by Beug (1961) does not occur in Northwest Europe (Graebner 1900), so this taxon has not been studied.

In Europe *T. shuttleworthii* seems to be limited to the valleys of the Alps and the Pyrenees, but the species is sometimes also found in South Germany. The pollen grains are similar in ornamentation to *Typha latifolia*.

Typha minima is more extensive in our area, and the pollen grains of this species seem to differ slightly from those of the other two species. The ornamentation is more distinctly rugulate and the muri are more interrupted. These muri are usually simplicolumellate and the columellae are not distinct in L—O analysis. In most grains the nexine was distinctly thicker than in the sexine, particularly in the porus area. The ornamentation of the undivided grains of the tetrad is very similar to that of the grains of *Typha angustifolia*.

PLATE DESCRIPTIONS (all Plates × 2000, except as otherwise stated)

PLATE I (p.*82*)

Sparganium erectum L. ssp. *erectum* (Gottenbos s.n.)
1. Ornamentation at high focus.
2. Ornamentation at low focus; columellae rather indistinct.
3. Ectoporus.
4. Scanning electron micrograph; ornamentation (× 10,000).
5. Scanning electron micrograph; overall shape.
Sparganium erectum L. ssp. *neglectum* (Beeby) Schinz et Thellung (Porta s.n.)
6. Ectoporus.

PLATE II (p.*83*)

Sparganium erectum L. ssp. *neglectum* (Beeby) Schinz et Thellung (Porta s.n.)
1. Ornamentation at high focus.
2. Ornamentation at low focus; muri thick, dupli- to polycolumellate; columellae distinct.
Sparganium emersum Rehman (Botanisch Museum No. 88587A)
3. Ornamentation at high focus.
4. Ornamentation at low focus; muri thin, usually simplicolumellate.
5. Optical cross-section; outline.
6. Scanning electron micrograph; overall shape.
Sparganium angustifolium Michaux (Van der Voo 59619)
7. Ornamentation at high focus.
8. Ornamentation at low focus; lumina angular, very irregular, often elongated.

PLATE III (p.*84*)

Sparganium friesii Beurling (Karo 488)
1. Ornamentation at high focus.
2. Ornamentation at low focus.
Sparganium angustifolium Michaux (Bouchard s.n.)
3. Ectoporus.
4. Optical cross-section; outline.
Sparganium minimum Wallroth (Lanjouw s.n.: 5,6,8; Kirk 1144: 7)
5. Ornamentation at high focus.
6. Ornamentation at low focus; muri thick, often interrupted.
7. Optical cross-section; outline.
8. Ectoporus.

PLATE IV (p.*85*)

Sparganium glomeratum Laestiboudois (Hjelt s.n.)
1. Ornamentation at high focus.
2. Ornamentation at low focus; muri thick, often interrupted.
Sparganium minimum Wallroth (Lanjouw s.n.)
3. Scanning electron micrograph; outline and ectoaperture.
4. Scanning electron micrograph; ornamentation ($\times$ 10,000).
Typha angustifolia L. (Van Ooststroom 24690)
5. Scanning electron micrograph; ornamentation and ectoporus ($\times$ 10,000).
6. Scanning electron micrograph; overall shape and ectoaperture.

PLATE V (p.*86*)

Typha latifolia L. (Behrendsen 2861)
1. Optical cross-section; outline.
2. Ornamentation at high focus.
Typha angustifolia L. (Van Ooststroom 24690)
3. Ornamentation at low focus.
4. Ornamentation at high focus.
5. Scanning electron micrograph; ornamentation ($\times$ 10,000).

PLATE VI (p.*87*)

Typha latifolia L. (Behrendsen 2861)
1. Optical cross-section; outline.
Typha minima Funk (Behrendsen s.n.)
2. Scanning electron micrograph; overall shape ($\times$ 1,000).
3. Scanning electron micrograph; ornamentation ($\times$ 10,000).
Typha latifolia L. (Behrendsen 2861)
4. Ornamentation at high focus; connection between two monads at high focus.
5. Ornamentation at low focus; several connections at high and low focus.

PLATE I

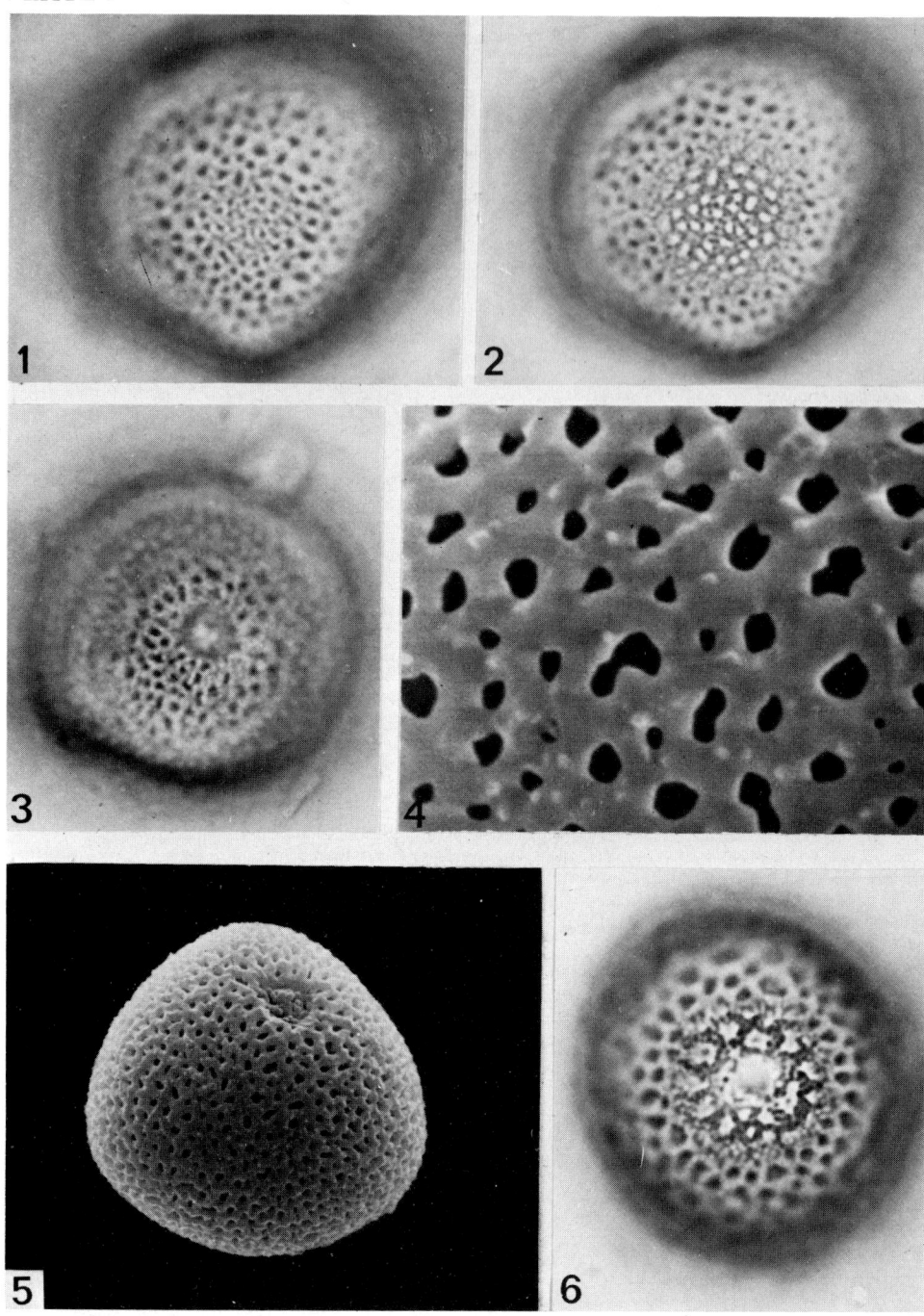

PLATE II

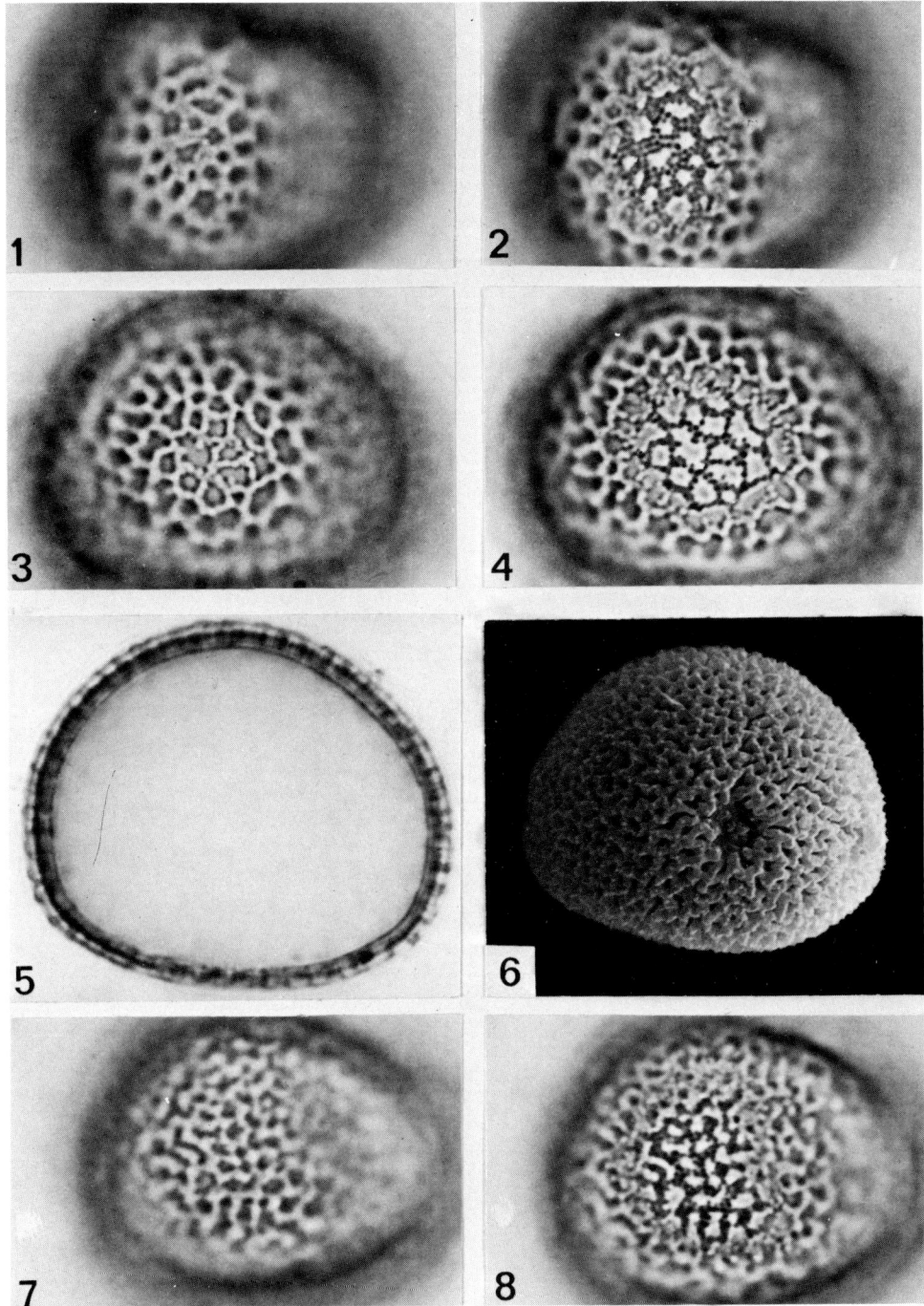

PLATE III

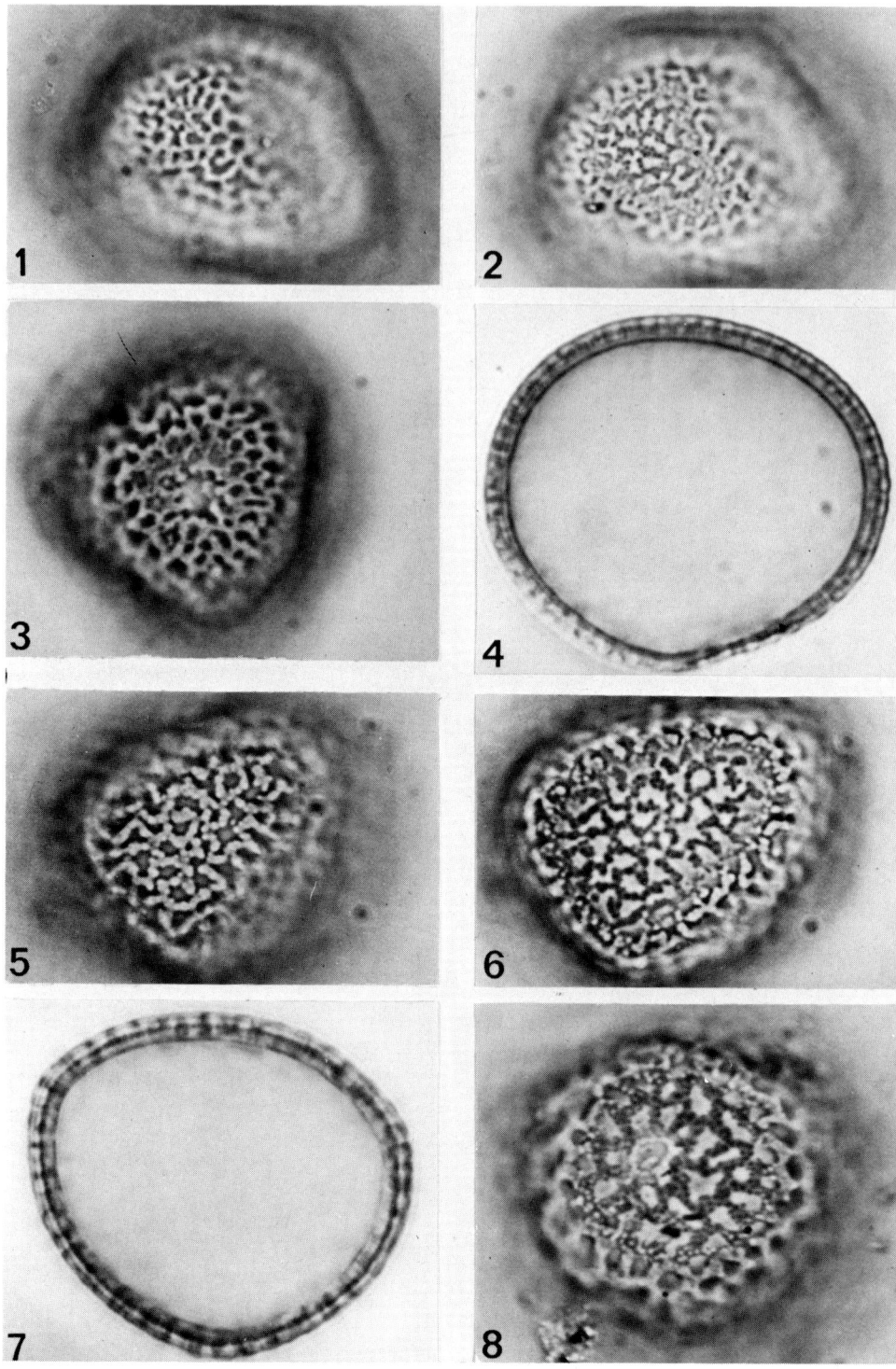

PLATE IV

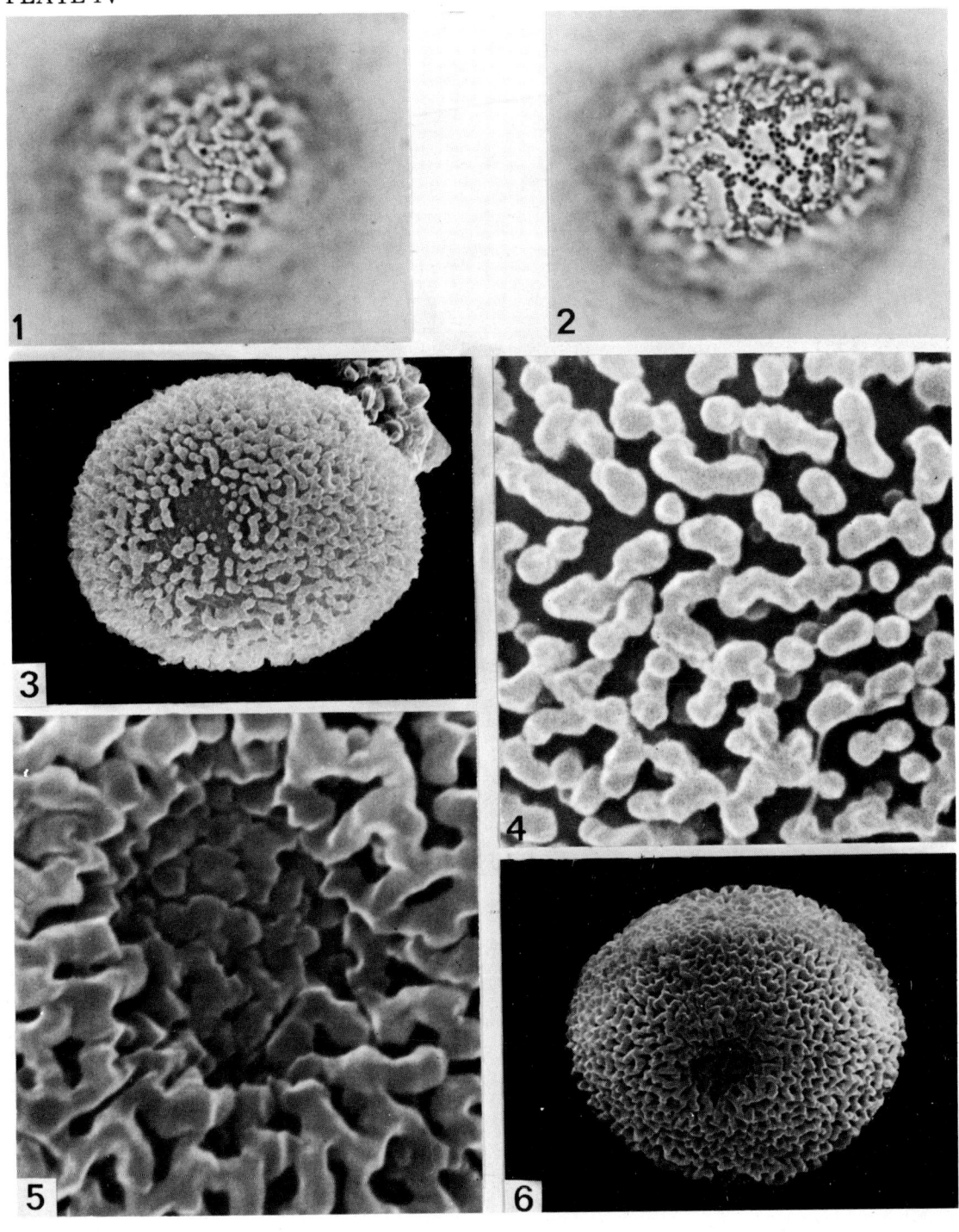

PLATE V

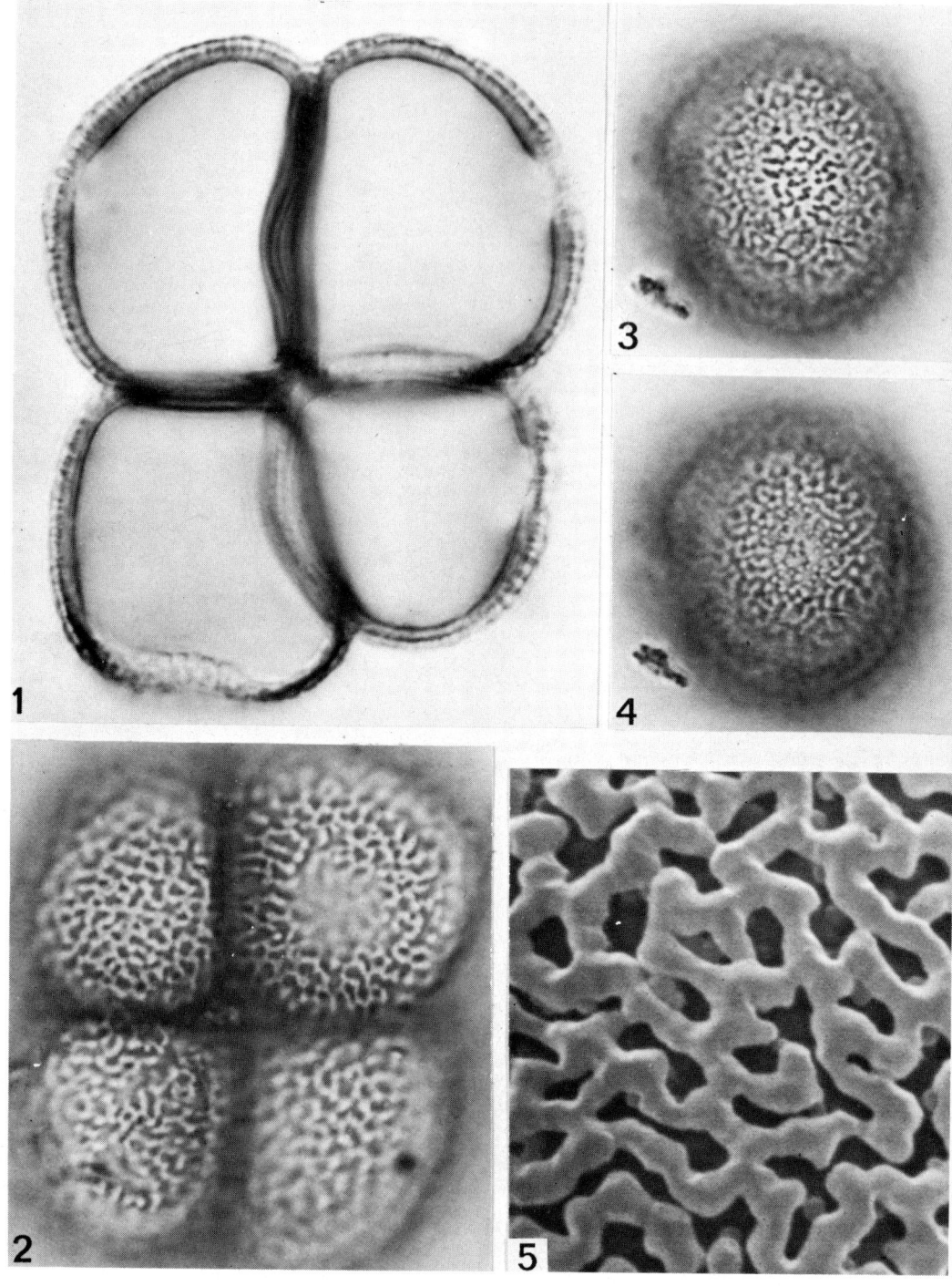

PLATE VI

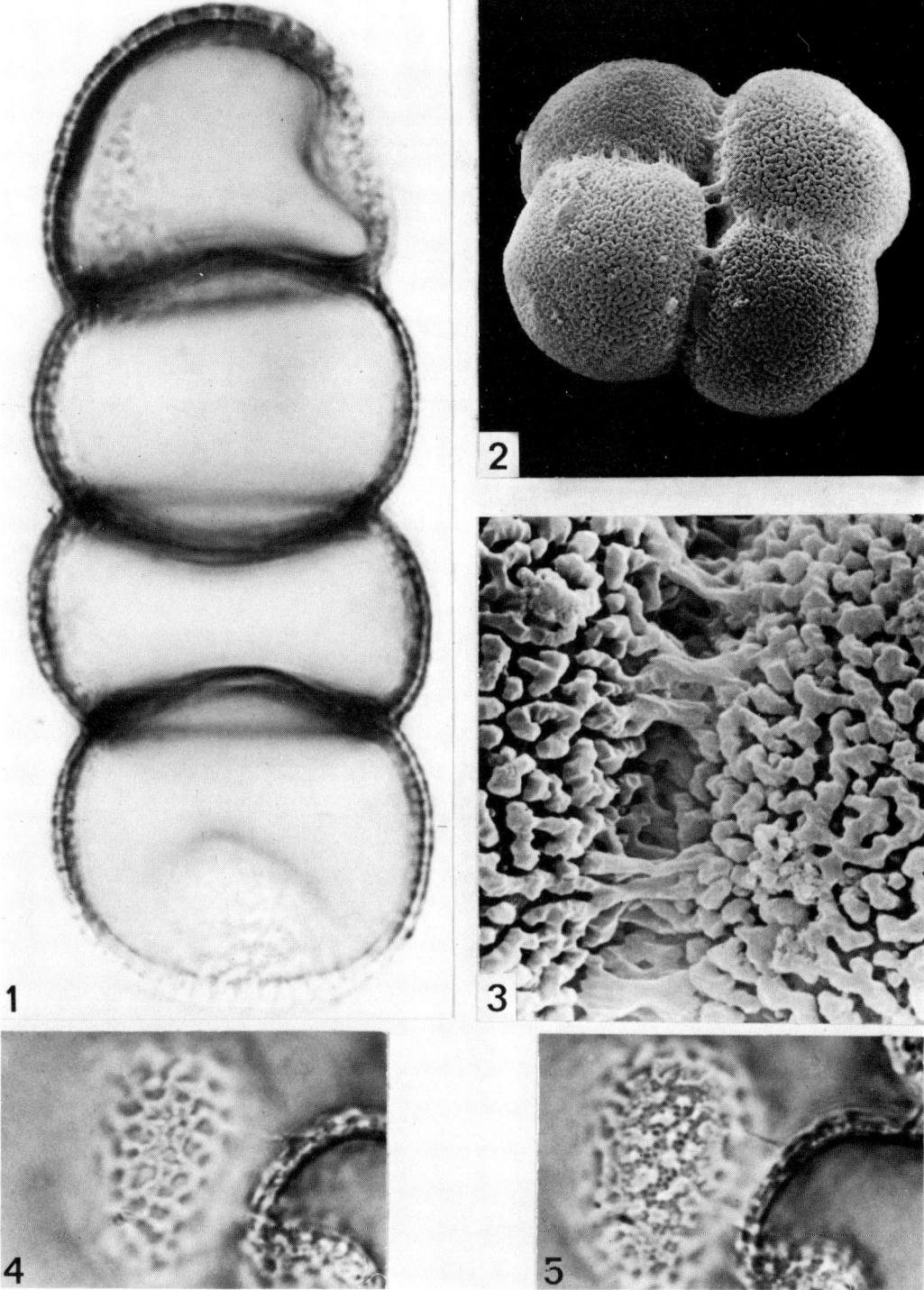

REFERENCES

Beug, H.-J., 1961. Leitfaden der Pollenbestimmung. Lieferung 1. Gustav Fischer Verlag, Stuttgart, 63 pp.

Cook, C. D. K., 1961. Sparganium in Britain. Watsonia, 5: 1—10.

Erdtman, G., 1943. An introduction to Pollen Analysis. Chronica Botanica, Waltham, Mass., 239 pp.

Erdtman, G., 1952. Pollen Morphology and Plant Taxonomy — Angiosperms. Almqvist and Wiksell, Stockholm, 1st ed., 539 pp.

Erdtman, G., Berglund, B. and Praglowski, J., 1961. An Introduction to a Scandinavian Pollen Flora, I. Almqvist and Wiksell, Stockholm, 92 pp.

Erdtman, G., Praglowski, J. and Nilsson, S., 1963. An Introduction to a Scandinavian Pollen Flora, II. Almqvist and Wiksell, Stockholm, 89 pp.

Faegri, K. and Iversen, J., 1950. Textbook of Modern Pollen Analysis. Munksgaard, Copenhagen, 169 pp.

Faegri, K. and Iversen, J., 1964. Textbook of Pollen Analysis. Munksgaard, Copenhagen, 2nd ed., 237 pp.

Graebner, P., 1900. In: A. Engler, Das Pflanzenriech, IV. 8. Typhaceae.

Van Ooststroom, S. J. and Reichgelt, Th. J., 1964. Flora Neerlandica, 1(6): 14. Sparganiaceae 227—238; 15. Typhaceae 239—242.

Visset, L., 1971. Quelques pollens actuelles en microscopie électronique à balayage. C. R. Acad. Sci. Paris, Ser. D., 273: 946—948.

Wodehouse, R. P., 1932. Tertiary pollen — I. Pollen of the living representatives of the Green River Flora. Bull. Torrey Bot. Club, 59: 313—340.

Wodehouse, R. P., 1935. Pollen grains. McGraw-Hill Book, New York and London, 574 pp.

The Northwest European Pollen Flora, 6

GENTIANACEAE

W. PUNT and W. NIENHUIS

Laboratory of Palaeobotany and Palynology, State University, Utrecht (The Netherlands)

LITERATURE

Andersen (1961), Erdtman (1943, 1952, 1969), Erdtman et al. (1961), Faegri and Iversen (1964), Kuprianova and Alyoshina (1972), Nilsson (1964, 1967a, b), Tarnavaschi and Serbanescu-Jitariu (1962).

INTRODUCTION

Pollen grains of the Gentianaceae have attracted the attention of botanists for many years. Already by 1895, Gilg had based parts of his key to the genera of the Gentianaceae on differences in pollen morphology. Subsequently Andersen (1961), Erdtman et al. (1961), Nilsson (1964, 1967a), Kuprianova and Alyoshina (1972) and Tarnavaschi and Serbanescu-Jitariu (1962) described pollen grains in more detail. The publications of Nilsson (1964, 1967a, b) are particularly interesting. His extensive work gives a good picture of the variability of pollen grains in the Gentianaceae—Gentianinae and it will be discussed in the comments on the pollen types. The comments of Faegri and Iversen in their Key to the N.W. European Pollen Types (1964) and of Erdtman (1952) are of lesser importance but nonetheless valuable. Erdtman, in the latter publication, pointed to some of the most interesting features in the Gentianaceae such as: "... pattern as a rule more delicate towards the colpi and margins of ora sometimes incrassate (and provided with ± fragmentary sexinous(?) elements)".

Comment on the previous descriptions by various authors will be given under the comments of the pollen types.

The genus is represented in Northwest Europe by seven genera of which *Gentiana, Gentianella* and *Centaurium* are by far the largest. The other four genera (*Cicendia, Blackstonia, Swertia* and *Lomatogonium*) have only one or two species in the area. The taxonomy and selection of species follows that of Flora Europaea (Tutin and Melderis, 1972).

TERMINOLOGY

H-endoaperture: A composite endoaperture, consisting of a central part which connects two lateral, longitudinally elongated legs, thus forming an H.

In the Gentianaceae the central part is rather complex. It consists of a more or less circular endoporus with two small and narrow lateral extensions which form the connection with the legs of the H. These legs may be short or long and are usually narrow, although sometimes rather broad, and then convex. The legs are best seen in the optical cross-section of the polar view when they appear as two gaps in the nexine immediately next to the deeply sunken colpus margins (Plate I, 7).

Striate: Reitsma (1970) gave the following definition: Pollen grain provided with a regular pattern of approximately parallel lumina and muri. Nilsson (1967a) defined the term more precisely: the fundamental, striate pattern is formed by ± parallel lirae separated by grooves (. . .). In this paper the basic concept of both authors is followed, but with the difference that the term lirae is replaced by muri and lumina by grooves. Our definition therefore runs: a pattern of ± parallel muri separated by grooves.

Striato-reticulate (Erdtman, 1952): Erdtman suggested the following definition: a sculptural pattern ± intermediate between striate and reticulate. In general this definition seems to give a good interpretation of the ornamentation occurring in the pollen grains of several taxa of the Gentianaceae. On the other hand, the term rugulate in the sense of Faegri and Iversen (1964) is not a striato-reticulate pattern although rugulate patterns are also considered intermediate between striate and reticulate. So Erdtman's concept is far too broad. Nilsson (1967a) also recognized this problem and commenting on Erdtman's term he tried to narrow the concept by adding: as long as the parallel arrangement of the lirae (muri) dominates at highest focus, the pattern is referred to as striate, whereas the "raising" of "bridges", interconnections, etc., *to the same level* (our italics) as the main lirae is regarded as striato-reticulation. By giving this comment, Nilsson excludes ornamentations covered by the term rugulate. In this paper the Nilsson (1967a) interpretation of the striato-reticulation is followed.

Retipilate (Erdtman, 1952): with a reticuloid pattern with pila instead of muri.

SPECIMENS EXAMINED

Blackstonia perfoliata (L.) Hudson ssp. *perfoliata* — France: Van Heerdt s.n.(U)., Van Loon 106(U); Ireland: Biol. Exc. 1969—134 (U).
Blackstonia perfoliata (L.) Hudson ssp. *serotina* (Koch ex Reichenbach) Vollmann — The Netherlands: Lindeman s.n. (U), Mees s.n. (U), Biol. Exc. 1916 (U).
Centaurium erythraea Rafn — Austria: Kramer and Westra 3850 (U); France: Baayens and Van der Ploeg s.n. (U); Ireland: Kertland s.n. (U), Biol. Exc. 1969—112 (U); Great Britain: Kramer 3724 (U), Sloet 4 (U).
Centaurium littorale (Turner) Gilmour — Austria: Kramer and Westra 3809 (U); Great Britain: Richards V-C.52 (U); The Netherlands: Borrias and De Bruyn s.n. (U).
Centaurium maritimum (L.) Fritsch — France: Reynders 1343 (L), (Corsica) Biol. Exc. 1965—623 (U); Italy (Elba): Kramer and Westra 3351 (U).
Centaurium pulchellum (Swartz) Druce — Denmark: Vogelpoel 71127 (U); The Netherlands: Van der Aa s.n. (U), Mennega s.n., Anno 1960 (U), Mennega (fresh material), Anno 1962.

Centaurium scilloides (L. fils) Sampaio — Portugal: De Smidt P102 (U); Spain: Merino s.n. (U), Biol. Exc. 1966—716 (U); The Netherlands: Boom B.1603 (cultivated) (U).
Centaurium spicatum (L.) Fritsch — France: Borders s.n. (L), s.c., s.n., Montpellier (U); Spain: Perez s.n. (U).
Centaurium tenuiflorum (Hoffmannsegg et Link) Fritsch — Great Britain: Groves and Tongen s.n. (U); France: Van Heerdt and Kramer 077 (U), Biol. Exc. 1954—479 (U).
Cicendia filiformis (L.) Delarbre — Ireland: Biol. Exc. 1969—164; Spain: Biol. Exc. 1957—903 (U); The Netherlands: Borrias and De Bruyn s.n. (U), Swart 2052 (U).
Gentiana cruciata L. — France: Biol. Exc. 1953—729 (U); The Netherlands: Van Royen and Van Egmond 2712 (U).
Gentiana lutea L. — Andorra: Biol. Exc. 1967—905 (U); France: Ten Berge 563 (U), Hekking 4000 (U), Biol. Exc. 1953—132 (U).
Gentiana nivalis L. — France: Biol. Exc. 1960—200 (U), Biol. Exc. 1966—2210 (U); Greenland: Daniels, Hooft and De Molenaar 182 (U); Iceland: Oosterveld 01071 (U).
Gentiana pneumonanthe L. — Austria: Van Leeuwen Reynvaan 6495 (U); Sweden: Hekking 3411 (U); The Netherlands: Punt (fresh material), anno 1961.
Gentiana purpurea L. — France: Biol. Exc. 1949—153 (U); Norway: Groet and Sloet 113 (U).
Gentiana verna L. — France: Biol. Exc. 1960—159 (U); Ireland: Biol. Exc. 1962—2558 (U); Switzerland: Biol. Exc. 1965—1400 (U).
Gentianella amarella (L.) Börner — Great Britain: Raven and Cannon 16489(BM), De Smidt 460 (U); The Netherlands: Brummer-De Vries s.n. (U).
Gentianella amarella (L.) Börner ssp. *septentrionalis* (Druce) Pritchard — Iceland: Ferman and Hille 54 (U).
Gentianella anglica (Pugsley) Warburg — Great Britain: Bradley s.n. (BM).
Gentianella aurea (L.) H. Smith — Norway: Hessel 1166 (U), Biol. Exc. 1965—2462 (U), Biol. Exc. 1965—2592 (U); Iceland: Ferman and Hille 122 (U).
Gentianella campestris (L.) Börner ssp. *baltica* (Murbeck) Vollmann — Denmark: Vogelpoel 71124 (U); The Netherlands: Vooren s.n. (U).
Gentianella campestris (L.) Börner ssp. *campestris* — Andorra: Biol. Exc. 1967—1134 (U); Austria: Van Soest 16873 (U); France: Eyma and Lanjouw s.n. (U); Iceland: Hartmann s.n. (U); Italy: Biol. Exc. 1960—672 (U); Norway: Arendsen-Hein s.n. (U).
Gentianella ciliata (L.) Borkhausen — France: Berg 14 (U); Germany, Behrendsen s.n. (U), Van der Aa s.n. (U).
Gentianella detonsa (Rottboell) G. Don — Norway: Oosterveld 0246 (U), Biol. Exc. 1965—2465 (U), Biol. Exc. 1965—2604 (U).
Gentianella germanica (Willdenow) Warburg — Belgium: Lindeman s.n. (U); France: Berg 15 (U); The Netherlands: Dijkstra 1129 (U).
Gentianella tenella (Rottboell) Börner — France: Biol. Exc. 1961—193 (U); Norway: Biol. Exc. 1965—245 (U); Sweden: Van Royen 2355 (U).
Gentianella uliginosa (Willdenow) Börner — Great Britain: De Smidt 460 (U); Norway: Draayer s.n. (U); The Netherlands: Den Hartog s.n. (U).
Lomatogonium carinthiacum (Wulfen) Reichenbach — Austria: Behrendsen s.n. (U); Switzerland: Braun-Blanquet 856 (U).
Lomatogonium rotatum (L.) Fries ex Fernald — Canada: Hooker s.n. (U); Iceland: Van der Veen s.n. (U).
Swertia perennis L. — Austria: Biol. Exc. 1963—4388 (U); France: Van Dam s.n. (U), Biol. Exc. 1953—703 (U).

KEY TO THE POLLEN TYPES

1.a. Colpate, colpi rather short; ornamentation reticulate, lumina not decreasing in size towards colpi . . . *Gentianella detonsa* type (p. *100*)
  b. Colporate, colpi long; ornamentation reticulate, striato-reticulate, tectate or striate, lumina decreasing in size towards the colpi 2

2.a. Mesocolpium tectate, striate to faintly striate; grooves between muri occasionally perforate, but if so, lumina narrower than muri 3
  b. Mesocolpium reticulate, microreticulate, often striato-reticulate or ± striate; lumina wider than muri 4
3.a. Poles acute in equatorial view; endoaperture margins smooth or with a few coarse granules; P/E ratio > 1; muri parallel to colpus . . . . .
  Gentiana pneumonanthe type (p. *96*)
  b. Poles obtuse in equatorial view; endoaperture margins coarsely granular; P/E ratio usually < 1, sometimes a little larger than 1; muri abruptly changing direction, often transverse to colpus . . . .
  Centaurium pulchellum type (p. *93*)
4.a. Endoaperture broadly elliptic, lalongate, ends obtuse; bridge often present. . . . . *Lomatogonium rotatum* type (p. *102*)
  b. Endoaperture ± circular or H-shaped, ends often with acute lateral extensions; no bridge 5
5.a. Lumina in mesocolpium wider than 1 µm, decreasing in size towards apocolpium . . . . *Gentianella campestris* type (p. *97*)
  b. Lumina ca. 1 µm or less in mesocolpium and apocolpium, no gradation in size 6
6.a. Ornamentation in mesocolpium striato-rugulate or striato-reticulate; muri more distinct near colpus; grains small (P < 25 µm) . . . . .
  Cicendia filiformis type (p. *95*)
  b. Ornamentation in mesocolpium microreticulate, finely reticulate or faintly striato-reticulate, but muri never prominent; grain size variable, P usually > 25 µm 7
7.a. Costae distinct; H-endoaperture indistinct; margins central part thick and smooth; colpus membrane smooth; grains usually 4-colporate, less frequently 3-colporate . . . . . *Blackstonia perfoliata* type (p. *92*)
  b. Costae absent or faint; endoaperture a porus with thin indistinct margins often obscured by coarse granules on colpus membrane; grains 3-colporate . . . . *Gentianella tenella* type (p. *100*)

(N.B.: Pollen grains of the well known species *Gentiana clusii* Perrier et Songeon and *Gentiana acaulis* L. are not included in this key. They occur in high mountains and as alpine plants they fall beyond the scope of this flora. A preliminary study of their pollen grains showed that they fall in the *Gentianella campestris* type (p. *97*).

Blackstonia perfoliata type (Plate I, 1—7)

Pollen class: Usually 4-colporate, either zonocolporate or pantocolporate, sometimes 3-zonocolporate, rarely syncolporate or 5-colporate.
P/E ratio: Subtransverse to adequate, less frequently suberect.
Apertures: Ectoaperture — colpus, long to very long, rather broad or broad and deeply sunken; margins distinct, straight, ends acute to slightly obtuse; colpus membrane smooth or with a few granules near the endoaperture margin; no costae. Endoaperture — H-endoaperture; central part circular or

nearly so, slightly lolongate to slightly lalongate; margins distinct, but irregular, usually smooth, sometimes with a few coarse granules; lateral extensions small and narrow; legs narrow and rather short, extending about halfway to the poles; distinct costae around the central part of the H-endoaperture.
Exine: Thick. Sexine far thicker than nexine; sexine 1 distinctly thicker than sexine 2. Sexine 1 consisting of long columellae; sexine 2 semitectate.
Ornamentation: Partly striate, partly reticulate. Distinct striae on a tectum occur along the colpus edge and extend towards the poles. The striation ceases abruptly towards the mesocolpium where the ornamentation is more or less reticulate or microreticulate. Muri narrow, fused only in the upper part, simplicolumellate. Lumina distinctly decreasing in size towards the colpi, but not towards the apocolpium. Columellae more or less circular in outline at low focus, sometimes elliptic or subangular.
Outlines: Equatorial view — elliptic or circular. Polar view — usually more or less circular, sometimes slightly angular; if angular sides convex and apertures in the obtuse angles.
Measurements: Glycerine jelly — P 21—30 μm; E 20—32 μm; P/E ratio 0.87—1.14; exine up to 4 μm; diameter of central part up to 5 μm; lumina ca. 1 μm, sometimes up to 2 μm. Silicone oil — P 24—32 μm; E 23—30 μm; P/E ratio 0.87—1.14.
Species: Blackstonia perfoliata ssp. *perfoliata, B. perfoliata* ssp. *serotina*

Comments

It was not possible to separate the two subspecies of *Blackstonia perfoliata* using pollen morphology.

Characters which differentiate this type from the closely related *Cicendia filiformis* type are: (1) usually 4-panto- or zonocolporate; (2) usually subtransverse or adequate; (3) striate, tectate near the colpus and reticulate in the middle of the mesocolpium; (4) irregularly thickened margin of the central part of the H-endoaperture.

The present type also resembles the *Centaurium pulchellum* type, but it can be distinguished by: (1) striate, tectate near the colpus and reticulate in the middle of the mesocolpium; (2) margin of the central part of the H-endoaperture not raised and more or less smooth; (3) sexine always distinctly thicker than nexine. This last character is, however, not conclusive, as some *Centaurium* species also have the sexine thicker than the nexine.

Centaurium pulchellum type (Plate II, 1—7; Plate II, 1, 2)

Pollen class: 3-Zonocolporate. Sometimes 4-pantocolporate, rarely syncolporate.
P/E ratio: Usually semitransverse to subtransverse, sometimes adequate or suberect *(Centaurium spicatum)*.
Apertures: Ectoaperture — colpus, long to very long, rather broad or broad and deeply sunken; margins distinct, straight to slightly irregular; ends acute

or obtuse; colpus membrane smooth with the exception of a few coarse granules near the margin of the endoaperture. Endoaperture — H-endoaperture or porus. If H-endoaperture, central part more or less circular, margins distinct, but irregular, usually granular and slightly raised; extensions small, narrow, often indistinct; legs narrow and short. If porus, features the same as the central part of the H-endoaperture. Distinct costae around central part or porus. Margo sometimes present.

Exine: Rather thick. Sexine about as thick as nexine or a little thicker; sexine 1 usually thicker than sexine 2. Sexine 1 consists of distinct columellae; sexine 2 is tectate.

Ornamentation: Tectate, supra-striate. Striae fine; muri narrow, often anastomosing; the direction of the parallel striae may change abruptly from one side to the other, they often meet in the middle of the mesocolpium under angles (Plate II, 2); grooves between the striae perforated, perforations may be very small (*Centaurium tenuiflorum*) or wide (*C. scilloides*). Columellae more or less circular to angular in outline at low focus.

Outlines: Equatorial view — elliptic to circular. Polar view — usually more or less circular, sometimes slightly angular; if angular, sides distinctly convex and apertures in the obtuse angles.

Measurements: Glycerine jelly — P 21—33 µm; E 20—33 µm; P/E ratio 0.86—1.18; exine less than 0.5 µm, sometimes up to 1 µm. Silicone oil — P 20—33 µm; E 20—35 µm; P/E ratio 0.88—1.15.

Species: Centaurium erythraea, C. littorale, C. maritimum, C. pulchellum, C. scilloides, C. spicatum, C. tenuiflorum

Comments

The *Centaurium pulchellum* type is similar in many ways to the *Blackstonia perfoliata* type. The main difference is in the ornamentation which is tectate and striate in the *Centaurium pulchellum* type but reticulate to microreticulate in the mesocolpia of the *Blackstonia perfoliata* type.

It is difficult to separate the *Centaurium* species from each other. There are a number of differential characters, but the differences are so indistinct and inconstant that it would be misleading to provide a key. The species differ to a certain extent in the following characters: (1) Presence or absence of an H-endoaperture. (2) The presence of connections between the central part and the legs of the endoaperture. (3) The width of the perforations in the tectum. (4) The sexine/nexine ratio. (5) The presence of a margo.

Ad 1. There is a distinct H-endoaperture in *C. pulchellum* and *C. erythraea*, a less distinct one in *C. littorale, C. spicatum* and *C. maritimum* and it seems to be lacking in *C. scilloides* and *C. tenuiflorum*. Ad 2. Distinct connections have been seen in *C. scilloides* and *C. spicatum*. This character is less distinct in the species *C. littorale, C. pulchellum, C. tenuiflorum* and *C. maritimum* and seems to be completely absent in *C. erythraea*. Ad 3. The perforations are widest in *C. scilloides* and *C. erythraea*. All other species have small to extremely small perforations. Ad 4. The sexine/nexine ratio is usually about 1, but in *C. scilloides* and *C. maritimum* the sexine is thicker than the nexine.

Ad 5. The presence of a margo is often difficult to determine. It is more or less distinct in *C. littorale, C. tenuiflorum* and *C. maritimum*, but not necessarily in every grain of a sample.

Because of all these uncertainties the authors are reluctant to provide a key to the species using pollen characters. A detailed study might produce a reliable means of separating but then such an investigation is outside the scope of this project.

Cicendia filiformis type (Plate III, 3—7)

Pollen class: 3-Zonocolporate, rarely 4-zonocolporate with two pairs of crossing colpi.
P/E ratio: Usually suberect to semi-erect, rarely adequate or subtransverse.
Apertures: Ectoaperture — colpus, long, broad, deeply sunken; margins distinct, usually straight to slightly irregular; ends acute; colpus membrane smooth, but sometimes a few coarse granules near the endoaperture edge; no costae. Endoaperture — H-endoaperture, indistinct; central part usually more or less circular, occasionally lolongate, margins distinct, narrow, not elevated, irregular; lateral extensions faint, often indistinct; legs very narrow, rather short, extending about halfway to the poles; faint costae. H-endoaperture seen best in optical section of the polar view where it appears as two narrow gaps bordering the ectocolpus.
Exine: Thick, particularly in mesocolpium and apocolpium. Sexine distinctly thicker than nexine; even in thickness in mesocolpium as well as in apocolpium. Sexine 1 consists of distinct, long columellae; sexine 2 is semitectate or tectate.
Ornamentation: Microreticulate to tectate; if tectate irregularly supra-striate to supra-rugulate. Striae short, running transversely to the colpus; striate and rugulate ornamentation occur in the same grain. Muri narrow; grooves bridged forming the lumina of a striato-reticulate pattern; lumina decreasing in size towards colpi, but not towards apocolpium. Columellae coarse at low focus, irregular in outline and inordinately arranged.
Outlines: Equatorial view — elliptic, rarely circular. Polar view — circular to more or less angular; if angular, with convex sides and obtuse angles; position of the apertures not clear.
Measurements: Glycerine jelly — P 19—26 μm; E 15—25 μm; P/E ratio 0.95—1.33; exine up to 3 μm; diameter central part up to 5 μm; apocolpium index 0.20—0.35; lumina of the reticulum less than 1 μm. Silicone oil — P 20—27 μm; E 20—24 μm; P/E ratio 0.91—1.23.
Species: Cicendia filiformis

Comments

The pollen grains of *Cicendia filiformis* are characterized by their small size, the irregular mesocolpial ornamentation which varies from striato-reticulate to more or less striato-rugulate with the pattern normally running transversely to the colpi.

This pollen type is not easily confused with any other in the Gentianaceae with exception of the *Blackstonia perfoliata* type. The latter is also striate next to the colpi, but is more distinctly reticulate to microreticulate in the mesocolpium. Any tendency to a striato-reticulate pattern is at the most slight. Both types have a more or less distinct H-endoaperture but they differ in the margins of the central part which are narrow and smooth in the *Cicendia filiformis* type but irregular and thick in the *Blackstonia perfoliata* type.

Gentiana pneumonanthe type (Plate IV, 1—7; Plate V, 1—6; Plate VI, 1, 2)

Pollen class: 3-Zonocolporate, rarely 4-zonocolporate.
P/E ratio: Usually suberect to semi-erect, rarely adequate or subtransverse.
Apertures: Ectoaperture — colpus, long to very long, rather broad, deeply sunken; margins distinct, straight to a little irregular; ends acute; colpus membrane smooth but with an uneven surface and sometimes with a few coarse granules near the endoaperture margin; no costae or short costae extending from the endoaperture towards the poles; apocolpium small. Endoaperture — porus, more or less circular in outline with two acute, lateral extensions which are as long as the height of the colpus margin in the equatorial plane; margins distinct; distinct costae.
Exine: Rather thick. Sexine about as thick as nexine or thicker. Sexine 1 about as thick as sexine 2 or thicker. Sexine 1 consists of thin, but distinct columellae; columellae often slightly higher and thicker at the apocolpium; sexine 2 is a tectate layer.
Ornamentation: Tectate, finely supra-striate. Muri thin, running parallel to the colpi, frequently anastomosing; grooves with numerous, very small perforations. Columellae circular in outline at low focus, irregularly distributed; usually slightly coarser at the apocolpium.
Outlines: Equatorial view — elliptic, sometimes subrhombic; either acute or obtuse towards the poles. Polar view — usually a little angular, sometimes more or less circular; if angular, sides convex with deeply sunken apertures in the very obtuse angles.
Measurements: Glycerine jelly — P 28—39 μm; E 25—34 μm; P/E ratio 1.04—1.33; exine 1.5—3 μm; endoporus diameter 3.5—7 μm; apocolpium index 0.10—0.35. Silicone oil — P 26—35 μm; E 23—32 μm; P/E ratio 0.94—1.35.
Species: Gentiana cruciata, G. lutea, G. pneumonanthe, G. purpurea

Key to the species
1.a. Sexine about as thick as nexine in mesocolpium and towards apocolpium; outline in equatorial view elliptic obtuse . . . . . *G. purpurea*
   b. Sexine thicker towards apocolpium where columellae are slightly coarser than in mesocolpium; in equatorial view variable in outline 2
2.a. Outline in equatorial view elliptic obtuse, never subrhombic *G. cruciata*

b. Outline in equatorial view elliptic, acute towards the poles; often slightly subrhombic if a colpus is located in optical section . . . . *G. lutea*
G. pneumonanthe

Comments

Nilsson (1967a), commenting the section *Coelanthe*, states that *Gentiana lutea* has more faintly striate pollen grains than the other species he studied (such as *G. purpurea*). *G. purpurea* itself is not described, but Nilsson mentions that its striation is similar to that of *G. pannonica* (not found in Northwest Europe) where "the striation is coarser than in *G. lutea*". Although this may be true, the differences in striation involved are so small that they cannot be used alone to give a reliable separation of *G. lutea* and *G. purpurea* or, indeed, between these and such species as *G. pneumonanthe* (Section *Pneumonanthe*) or *G. cruciata* (Section *Aptera*). In this key we have felt it better to rely on columellae height and equatorial view outline differences, but even these features are difficult to observe and only visible in perfectly formed pollen grains.

Andersen (1961) gives a short description of *G. pneumonanthe* and associates *C. asclepiadacea*, *G. lutea*, *G. cruciata* and *G. purpurea* with it, but he does not attempt any further differentiation.

Gentianella campestris type (Plate VI, 3—6; Plate VII, 1—6; Plate VIII, 1—5; Plate IX, 5, 6; Plate X, 1—6; Plate XI, 1—3)

Pollen class: 3-Zonocolporate, rarely 4-colporate, colpi crossed two by two.
P/E ratio: Usually suberect to semi-erect, less frequently adequate or subtransverse.
Apertures: Ectoaperture — colpus, long, broad, deeply sunken; margins distinct, smooth to slightly irregular; ends acute to obtuse; colpus membrane smooth or with very small granules and often a few coarse granules near the margins of the endoaperture; no costae. Endoaperture — porus, more or less circular, usually with distinct, acute, lateral extensions which are as long as the height of the colpus margin in the equatorial plane; margins distinct, usually irregular because of the coarse granules on the colpus membrane and often elevated; costae absent or faint.
Exine: Thick. Sexine usually distinctly thicker than nexine or sometimes about as thick; usually even in thickness throughout the mesocolpium and apocolpium, sometimes a little thicker at apocolpium (*Gentiana verna*), sometimes thinner (*Gentiana nivalis*). Sexine 1 about as thick as sexine 2 or slightly thicker. Sexine 1 consists of distinct columellae which are often rounded at the top, sometimes acute (*Gentianella ciliata*); sexine 2 semitectate.
Ornamentation: Reticulate or faintly to distinctly striato-reticulate in mesocolpium; in apocolpium microreticulate or finely reticulate. Muri relatively thick, often fused at top only but sometimes to at least halfway down, simplicolumellate, rarely duplicolumellate. Lumina irregular, angular, often widest at top, distinctly decreasing in size towards the colpi and towards the poles. Columellae varying in outline at low focus, either circular or angular.

Outlines: Equatorial view — elliptic, rarely circular, sometimes subrhombic; poles either acute or obtuse. Polar view — usually angular, sometimes more or less circular; sides convex and apertures in the obtuse angles.
Measurements: Glycerine jelly — P 27—57 µm; 24—57 µm; P/E ratio 0.96—1.38; exine 2—3.5 µm; endoaperture diameter 4—8 µm; apocolpium index 0.10—0.45; lumina in mesocolpium often ca. 1.5 µm, sometimes up to 3 µm, in apocolpium usually less than 1 µm, sometimes up to ca. 1 µm.
Silicone oil — P 27—45 µm; E 26—45 µm; P/E ratio 0.86—1.21.

	P(µm)	E(µm)	P/E	Lumina in mesocolpium (µm)
Gentianella campestris				
ssp. *campestris*	48—55	43—57	0.97(1.10)1.24	2
ssp. *baltica*	35—50	37—46	0.95(1.00)1.10	1.5
Gentianella anglica	46—57	38—55	1.02(1.09)1.24	2
Gentianella germanica	42—55	40—51	1.03(1.08)1.18	2
Gentianella uliginosa	40—50	35—45	0.91(1.04)1.14	2.5
Gentiana ciliata	27—43	24—41	1.01(1.05)1.13	2.5
Gentiana nivalis	37—49	29—46	1.08(1.18)1.25	3
Gentiana verna	39—48	31—40	1.04(1.24)1.38	3

Species: Gentiana ciliata, Gentiana verna, G. nivalis, Gentianella anglica. Gentianella campestris ssp. *baltica, Gentianella campestris* ssp. *campestris, G. germanica, G. uliginosa (Gentiana acaulis, Gentiana clusii)*

Key to the species and groups

1.a. Columellae at low focus ordinately arranged in a reticulate pattern . . 3
 b. Columellae at low focus inordinately arranged, not forming a reticulum 2
2.a. Muri fused for at least halfway their length, columellae short and indistinct . . . . *Gentianella uliginosa*
 b. Muri fused at top only, columellae long and distinct . . . .
 . . . . *Gentianella campestris* group
3.a. Endoaperture with distinct, acute lateral extensions; P/E ratio usually suberect to subtransverse; sexine about as thick as nexine, even in thickness throughout . . . . *Gentianella ciliata*
 b. Endoaperture without distinct, acute, lateral extensions; P/E ratio usually semi-erect, less frequently suberect or erect; sexine distinctly thicker than nexine, at apocolpium thicker or thinner. 4
4.a. Endoaperture a porus; sexine usually thicker at apocolpium; sexine 1 in apocolpium thicker than sexine 2; outline in polar view distinctly angular. *Gentiana verna*
 b. H-endoaperture (best seen in polar view as two gaps beside the colpus) sexine usually thinner at apocolpium than in mesocolpium; sexine 1 in apocolpium about as thick as sexine 2; outline in polar view more or less circular to slightly angular. . . . . *Gentiana nivalis*

Comments

Andersen (1961) recognized a *Gentianella amarella* type and he combined all *Gentianella* species (*Gentianella tenella* excepted) in this type with a number of *Gentiana* species such as *Gentiana nivalis* and *Gentiana verna*. In the present *Gentianella campestris* type the same *Gentiana* species have been combined with most of the *Gentianella* species with *Gentianella tenella* still kept separate. However, *Gentianella amarella* and *Gentianella aurea*, which were combined by Andersen with the other *Gentianella* species are associated here with *Gentianella tenella*. Andersen's type description is very short whereas the present one is more detailed and exclusive.

Nilsson (1967a) uses the taxonomic classification as basis for his description of the pollen grains of the Gentianinae. As a result *Gentiana nivalis* and *Gentiana verna* were described separately from the *Gentianella* species. Not surprisingly he states that "the pollen morphological data obtained during this study are sometimes too confusing to allow definite conclusions. This is to a great extent due to the fact that similar pollen morphological types appeared in entirely unrelated genera and sections". *Gentiana nivalis* and *Gentiana verna* were both described under the section *Cyclostigma* with no differences given between the taxa. The general description of the whole section, however, gives some of the most characteristic features of the two species (pollen grains subprolate to prolate; exine usually thinner at poles; sexine coarsely striate or striato-reticulate).

Nilsson (1967a) followed Andersen (1961) in combining *Gentianella amarella* with the other species of Section *Amarella* such as *Gentianella campestris* and *Gentianella uliginosa* (*Gentianella germanica* and *Gentianella anglica* were not studied). Although there is a close resemblance between the pollen grains of all the *Gentianella* species, it seems best to separate them into two types with the main difference being the reduction of lumina size near the apocolpium in *Gentianella amarella*, *G. aurea* and *Gentianella tenella*.

It is difficult to differentiate between the species of the *Gentianella campestris* group. The pollen grains of *Gentianella germanica* show the coarsest striato-reticulate pattern with relatively thick muri distinctly fused at the top. *Gentianella anglica* and *Gentianella campestris* ssp. *campestris* also have thick muri which are distinctly fused, but in these species the ornamentation is less prominently striato-reticulate. The pollen grains of *Gentianella campestris* ssp. *baltica* are very much like those of *Gentianella campestris* ssp. *campestris*, but in the former the muri are only loosely fused at the top and the striato-reticulate ornamentation is even less distinct. Since the differences between the taxa are so vague they should be placed in one group.

The differences between the *Gentianella campestris* group and the other species are clear and more constant, but the overall similarities in colpus features and in ornamentation makes them a cohesive type.

Gentiana clusii and *Gentiana acaulis*, although only superficially studied can be differentiated from the other species of this type by their distinct, coarse striato-reticulate pattern with sharp-edged muri. Other differential characters are: outline in equatorial view subrhombic, distinctly acute towards the pole; sexine raised in the equatorial plane near the endoaperture; P/E ratio suberect to semi-erect; size of the grains large (P 45—55 µm).

Gentianella detonsa type (Plate XI, 4—6; Plate XII, 1, 2)

Pollen class: 3—5-Zonocolpate, sometimes more or less pantocolpate; if 4-colpate, colpi often crossed two by two.
P/E ratio: Usually semitransverse or subtransverse, rarely adequate or suberect.
Apertures: Ectoaperture — colpus, rather short and narrow, not or only slightly sunken; margins distinct, irregular; ends acute; colpus membrane with some small granules or smooth; no costae. Endoaperture — absent.
Exine: Thick. Sexine about as thick as nexine or thicker. Sexine 1 about as thick as sexine 2 or thinner. Sexine 1 consists of low, distinct columellae; sexine 2 semitectate.
Ornamentation: Reticulate throughout. Muri relatively thin, simplicolumellate, at least halfway fused, interrupted in many pollen grains, but more so in some specimens than in others; lumina irregular in outline, varying in width, small and large lumina intermixed, not decreasing in size towards the colpi. Columellae more or less angular in outline, at low focus arranged in a reticulate pattern.
Outlines: Equatorial view — elliptic, rarely circular. Polar view — circular to a little angular; sides convex; apertures in the obtuse angles.
Measurements: Glycerine jelly — P 35—46 µm; E 35—46 µm; P/E ratio 0.87—1.04; exine 1.5—3 µm; apocolpium index larger than 0.40; lumina usually ca. 2 µm, up to 3 µm. Silicone oil — P 35—46 µm; E 36—45 µm; P/E ratio 0.87—1.10.
Species: Gentianella detonsa.

Comments

Nilsson (1967a) describing pollen grains of the section *Crossopetalum* which includes *Gentianella detonsa* as well as *Gentianella ciliata* states: "... the European species *G. ciliata* has pollen grains that deviate from the other species studied by being subprolate to prolate in shape and by having relatively small apocolpia and lolongate, well delimited ora". The most characteristic feature is the coarse reticulum which does not decrease in size towards the short colpus, but this is not referred to by Nilsson.

This is the most easily recognized type among the Gentianaceae of Western Europe and there are no problems in its identification.

Gentianella tenella type (Plate IX, 1—4; Plate XII, 3—6; Plate XIII, 1—8; Plate XIV, 1—7)

Pollen class: 3-Zonocolporate, rarely 4-colporate with two pairs of crossing colpi.
P/E ratio: Usually subtransverse to semi-erect, occasionally semitransverse.

Apertures: Ectoaperture — colpus, long, broad, deeply sunken; margins distinct, straight or a little irregular; ends acute or obtuse; colpus membrane smooth or with small, scattered granules (*Swertia perennis*), often some coarse granules near the margin of the endoaperture; no costae. Endoaperture — porus, outline more or less circular with or without two acute, lateral extensions; margins distinct, usually irregular because of some coarse granules, often slightly elevated, sometimes narrow and smooth (*Swertia perennis*); costae faint or absent.

Exine: Thick. Sexine as thick as nexine or thicker, of constant thickness throughout the grain. Sexine 1 about as thick as sexine 2 or thicker. Sexine 1 consists of short or long columellae; sexine 2 semitectate or rarely pertectate in some parts of the grain. Sexine sometimes slightly raised at the equator near the endoaperture.

Ornamentation: Finely striate to finely striato-reticulate in mesocolpium, microreticulate at apocolpium. Muri fused at top distally only, simplicolumellate; lumina varying in outline from more or less circular to irregularly angular, usually 1 μm or less, rarely larger, distinctly decreasing in size towards colpi, but not towards apocolpium. Columellae varying in outline at low focus, usually more or less angular, inordinately arranged.

Outlines: Equatorial view — elliptic to subrhombic with obtuse polar ends, sometimes circular. Polar view — circular or angular; sides convex, apertures in the obtuse angles.

Measurements: Glycerine jelly — P 24—51 μm; E 23—44 μm; P/E ratio 0.77—1.32; exine 1.5—4 μm; diameter of endoaperture 3—7 μm; apocolpium index 0.10—0.50; lumina usually smaller than 1 μm, sometimes up to ca. 1 μm. Silicone oil — P 24—47 μm; E 26—45 μm; P/E ratio 0.74—1.26.

Species: Gentianella amarella ssp. *amarella, Gentianella amarella* ssp. *septentrionalis, Gentianella aurea, Gentianella tenella, Swertia perennis*

Key to the species

1.a. Grains large (P > 36 μm); outline in equatorial view often subrhombic; outline in polar view usually angular; P/E ratio often suberect, less frequently adequate or semi-erect, rarely subtransverse . . . . . .
Gentianella amarella
 b. Grains smaller (P < 35 μm); outline in equatorial view circular to elliptic; outline in polar view circular or angular; P/E ratio variable . . . 2
2.a. Sexine about as thick as nexine; columellae rather short and indistinct; ornamentation in mesocolpium more or less reticulate; outline in polar view more or less circular. . . . . *Gentianella tenella*
 b. Sexine distinctly thicker than nexine; columellae long and distinct; ornamentation in mesocolpium faintly striato-reticulate or faintly striate, outline in polar view angular 3
3.a. Colpus membrane usually with small, scattered granules; apocolpium index usually between 0.10—0.30; endoaperture margins narrow, distinct; sexine slightly elevated at equator near endoaperture. . . .
Swertia perennis

b. Colpus membrane smooth; apocolpium index usually between 0.25—0.45; endoaperture margins relatively thick and vague; sexine not elevated at equator near endoaperture . . . .*Gentianella aurea*

Comments

This pollen type is characterized by the microreticulate ornamentation which does not perceptibly decrease in size towards the poles. In addition the features of the colpus, endoporus and colpus membrane, together with the obtuse poles in equatorial view separate this type from others.

The species in this type can easily be differentiated with a high power objective. Sexine features are especially important as are differences in ornamentation.

The illustrations of *Gentianella aurea* and *Gentianella tenella* given by Nilsson (1967a) clearly show the main differences, viz:
(1) columellae long in *G. aurea* but short in *G. tenella*; (2) striato-reticulate ornamentation in *G. aurea* but microreticulate or retipilate in *G. tenella*.

Andersen (1961) illustrates the ornamentation in the mesocolpium of *Swertia perennis* and also shows an equatorial cross-section. His description is short and not particularly conclusive (3-colporate, psilate with coarse perforations or sometimes indistinctly rugulate—striate, columellae evenly scattered). Nilsson's description (1967a, b) is even shorter and less conclusive (the pollen grains are finely striate: ± smooth, with thick, densely scattered bacula, sexine pattern does not resemble that of the North American *Swertia* species). According to Nilsson there is little variation in pollen morphology between specimens, a feature which he also noted in the related *Lomatogonium rotatum*.

Swertia perennis can be separated from the *Gentianella* species in this type by the granules on the colpus membrane, the small apocolpium index and the sexine elevation at the equator. The features may not always be distinct, but the combination of characters is sufficient to separate *Swertia perennis* from *Gentianella*. Some other features, although less reliable, may be of some value; viz. colpi in *Swertia perennis* wide and deeply sunken, margins of the endoapertures distinct, but narrow and often smooth; exine relatively thin, usually between 2—3 μm (thicker in *Gentianella aurea* and *Gentianella amarella*, but similar in *Gentianella tenella*).

Lomatogonium rotatum type (Plate XV, 1—7)

Pollen class: 3-Zonocolporate, rarely 4-zonocolporate.
P/E ratio: Usually subtransverse or adequate, less frequently suberect.
Apertures: Ectoaperture — colpus, long, rather narrow, sunken; margins often rather indistinct because of the gradual transition from sexine ornament to colpus; ends obtuse, colpus membrane smooth to finely granulate; at equator sexine slightly raised and often bridging the endoaperture, fastigium absent; no costae. Endoaperture — porus, rather large, distinct lalongate, wider than colpus; margins distinct; outline broadly elliptic, truncate; ends obtuse; no costae.

Exine: Thick. Sexine about as thick as nexine or a little thicker. Sexine 1 about as thick as sexine 2; sexine 1 consists of short distinct columellae; sexine 2 semitectate.
Ornamentation: Finely striato-reticulate. Muri thin, fused at top only, simplicolumellate. Lumina distinct, but usually less than 1 μm wide and decreasing in size towards the colpi, but not towards the apocolpium. Columellae coarse, at low focus inordinately arranged, irregular in cross-section.
Outlines: Equatorial view — elliptic or circular. Polar view — more or less circular to slightly angular; sides convex, apertures in the obtuse angles.
Measurements: Glycerine jelly — P 22—34 μm; E 25—35 μm; P/E ratio 0.86—1.04; exine 2.5—4 μm; maximum breadth of the endoaperture up to 8 μm; apocolpium index 0.25—0.40; lumina usually less than 1 μm, sometimes up to ca. 1 μm. Silicone oil — P 25—32 μm; E 25—34 μm; P/E ratio 0.86—1.06.
Species: Lomatogonium carinthiacum, L. rotatum

Comments

Lomatogonium has been monographically treated by Nilsson (1964). According to him two main pollen types occur, one of them named the *Lomatogonium rotatum* type. Its description is as follows: "Pollen with lirae arranged in a ± striate pattern, no supratectal processes, puncta only occasionally present. Ora rather large, lalongate. The species within this group are pollen morphologically very similar to each other". Nilsson also places *Lomatogonium carinthiacum* in this type, a decision which has been confirmed in the present study. In the general description of *Lomatogonium* pollen Nilsson draws attention to the typical overlapping bridging of the sexine over the ora.

Lomatogonium rotatum is one of the species in the Gentianaceae with an appreciable proportion of variant pollen grains. Nilsson (1964) reported a high percentage of 4-colporate grains and a "sharp striation" in specimens from Iceland, whereas specimens from other countries were more constantly 3-colporate and less sharply striate.

ACKNOWLEDGEMENTS

The authors are very grateful to Miss M. R. Jones (British Museum, Natural History, London) for taking the SEM-micrographs of *Gentianella amarella* and also to Miss F. J. Blok (Centrum voor Submicroscopisch Onderzoek, Utrecht) for her assistance in taking all other SEM micrographs.

PLATE DESCRIPTIONS (all plates × 2000, except scanning electron micrographs)

PLATE I (p. *108*)

Blackstonia perfoliata (L.) Hudson (Biol. Exc. Ireland 1969-134)
1. Scanning electron micrograph; polar view (× 2000).
2. Scanning electron micrograph; ornamentation, distinctly striate along the colpus, more or less reticulate in the middle of the mesocolpium (× 5000).
3. H-endoaperture; rather short legs and distinct costae around the central part.
4. Ornamentation in mesocolpium at high focus; striato-reticulate pattern.
5. Ornamentation in mesocolpium at low focus; columellae thick, irregular and inordinately arranged.
6. Optical cross-section; outline in equatorial view, columellae distinctly higher at the poles.
7. Optical cross-section; outline in polar view, distinct two gaps near the colpus in the endexine.

PLATE II (p. *109*)

Centaurium pulchellum (Swartz) Druce (Vogelpoel 71127)
1. Scanning electron micrograph; equatorial view, colpus membrane with coarse granules near the endoaperture (× 3000).
2. Scanning electron micrograph; ornamentation in mesocolpium, striae meet under angles (× 6000).
Centaurium pulchellum (Swartz) Druce (Van der Aa s.n.)
3. Endoaperture at high focus; many coarse granules around the margin.
4. Endoaperture at lower focus.
5. H-endoaperture at lowest focus; distinct legs and small, narrow connections with the central part.
6. Ornamentation at high focus; striae meet under angles.
7. Ornamentation at lower focus; distinct perforations.
8. Ornamentation at lowest focus; columellae short and inordinately arranged.

PLATE III (p. *110*)

Centaurium erythraea Rafn (Baayens and Van der Ploeg s.n.)
1. Scanning electron micrograph; polar view, striae meet under angles (× 2500).
2. H-endoaperture; coarse granules around the margin of the central part.
Cicendia filiformis (L.) Delarbre (Swart 2052).
3. Scanning electron micrograph; equatorial view (× 2500).
4. Scanning electron micrograph; ornamentation in mesocolpium (× 5000).
5. Optical cross-section; outline in polar view, distinct gaps bordering the ectocolpus.
6. Ornamentation in mesocolpium at high focus; rugulate pattern.
7. Ornamentation in mesocolpium at low focus; columellae inordinately arranged.

PLATE IV (p. *111*)

Gentiana pneumonanthe L. (Van Leeuwen Reynvaan 6495)
1. Scanning electron micrograph; equatorial view (× 2000).
2. Scanning electron micrograph; ornamentation in mesocolpium, striate pattern (× 5000).
3. Optical cross-section; outline in equatorial view; sexine 2 distinctly thicker towards the poles, but not at the poles, elliptic acute in outline.

5. Ornamentation in apocolpium; columellae inordinately arranged.
Gentiana lutea L. (Hekking 4000)
4. Optical cross-section; outline in polar view.
6. Ornamentation in apocolpium at low focus.
7. Ornamentation in apocolpium at high focus.

PLATE V (p. *112*)

Gentiana cruciata L. (Van Royen 712)
1. Optical cross-section; outline in equatorial view, elliptic obtuse.
2. Colpus at low focus; margins of central part with costae, distinct acute lateral extensions.
Gentiana lutea L. (Hekking 4000)
3. Optical cross-section; outline in equatorial view; sexine 2 distinctly thicker towards the poles, outline elliptic acute.
4. Colpus and endoaperture.
5. Ornamentation in mesocolpium at high focus; faintly striate.
6. Ornamentation in mesocolpium at low focus; columellae circular in outline, inordinately arranged.

PLATE VI (p. *113*)

Gentiana purpurea L. (Groet and Sloet 113)
1. Optical cross-section; outline in polar view, more or less circular.
2. Optical cross-section; outline in equatorial view; sexine 2 not thicker towards the poles, obtuse elliptic in outline.
Gentianella campestris (L.) Börner ssp. *baltica* (Murbeck) Vollmer (Vogelpoel 71124)
3. Ornamentation in mesocolpium at high focus; capita fused at top only.
4. Ornamentation in mesocolpium at low focus; thick, irregular columellae, inordinately arranged.
Gentianella campestris (L.) Börner ssp. *campestris* (Biol. Exc. Andorra 1967-1134)
5. Ornamentation in mesocolpium at low focus; columellae low.
6. Ornamentation in mesocolpium at high focus; capita distinctly fused.

PLATE VII (p. *114*)

Gentianella ciliata (L.) Borkhausen (Van der Aa s.n.)
1. Scanning electron micrograph; outline in polar view (× 2000).
2. Scanning electron micrograph; ornamentation, sharp muri (× 5000).
3. Optical cross-section; outline in polar view circular.
4. Ornamentation in apocolpium at high focus; microreticulate.
5. Colpus and endoporus; two acute lateral extensions, coarse granules around the margin of the endoporus.
6. Optical cross-section; outline in equatorial view, about circular.

PLATE VIII (p. *115*)

Gentianella ciliata (L.) Borkhausen (Van der Aa s.n.)
1. Ornamentation in mesocolpium at high focus; lumina very irregular.
2. Ornamentation in mesocolpium at low focus; columellae circular in outline, arranged in a reticulate pattern.
Gentianella germanica (Willdenow) Warburg (Lindeman s.n.)

3. Optical cross-section; outline in polar view.
Gentianella germanica (Willdenow) Warburg (Dijkstra 1129)
4. Ornamentation in apocolpium; microreticulate.
Gentiana nivalis L. (Oosterveld 01071)
5. Apocolpium; lumina of reticulum distinctly decreasing in size towards the pole.

PLATE IX (p. *116*)

Gentianella amarella (L.) Börner ssp. *amarella* (Raven and Cannon 16489), see also Plate XII
1. Scanning electron micrograph; polar view (× 1850).
2. Scanning electron micrograph; ornamentation in mesocolpium, striato-reticulate pattern (× 4600).
3. Scanning electron micrograph; equatorial view (× 1850).
4. Scanning electron micrograph; colpus and endoporus, coarse granules near the margin of the endoaperture (× 4600).
Gentianella germanica (Willdenow) Warburg (Dijkstra 1299)
5. Ornamentation in mesocolpium at high focus; faint striato-reticulate pattern.
6. Ornamentation in mesocolpium at low focus; columellae thick, irregular in outline and inordinately arranged.

PLATE X (p. *117*)

Gentiana nivalis L. (Oosterveld 01071)
1. Optical cross-section; outline in equatorial view.
2. Ornamentation in mesocolpium at low focus; columellae arranged in a reticulate pattern.
3. Ornamentation in mesocolpium at high focus; capita distinctly fused.
Gentiana verna L. (Biol. Exc. France 1960-159)
4. Optical cross-section; outline in equatorial view.
5. Ornamentation in apocolpium at high focus; microreticulum.
6. Scanning electron micrograph; ornamentation in apocolpium (× 5000).

PLATE XI (p. *118*)

Gentiana verna L. (Biol. Exc. France 1960-159)
1. Ornamentation in mesocolpium at high focus.
2. Ornamentation in mesocolpium at low focus; columellae irregular in outline and arranged in a reticulate pattern.
3. Endoporus with coarse granules along the margin.
Gentianella detonsa (Rottboell) G. Don (Biol. Exc. Norway 1965-2604)
4. Optical cross-section; outline in polar view.
5. Colpus at low focus; colpus margin irregular.
6. Colpus at high focus; lumina not decreasing in size towards the colpi.

PLATE XII (p. *119*)

Gentianella detonsa (Rottboell) G. Don (Biol. Exc. Norway 1965-2604)
1. Ornamentation in mesocolpium at high focus; coarse reticulum, small and large lumina intermixed.
2. Ornamentation in mesocolpium at low focus; columellae irregular in size, more or less circular in outline, arranged in a reticulate pattern.
Gentianella amarella (L.) Börner ssp. *amarella* (Brummer — de Vries s.n.)

3. Optical cross-section; outline in equatorial view, subrhombic.
4. Ornamentation in mesocolpium at high focus; microreticulum.
5. Ornamentation in mesocolpium at low focus; columellae thick and irregular in outline, inordinately arranged.
6. Ornamentation in apocolpium at high focus; microreticulum.

PLATE XIII (p. *120*)

Gentianella aurea (L.) H. Smith (Hessel 1166)
1. Optical cross-section; outline in polar view, angular.
2. Ornamentation in mesocolpium at low focus; columellae thick and irregular in outline, inordinately arranged.
3. Ornamentation in apocolpium at high focus; microreticulum.

Gentianella tenella (Rottboell) Börner (Van Royen 2355)
4. Scanning electron micrograph; equatorial view (× 2000).

Gentianella tenella (Rottboell) Börner (Biol. Exc. France 1961—193).
5. Optical cross-section; outline in polar view more or less circular.
6. Optical cross-section; outline in equatorial view.
7. Ornamentation in mesocolpium at high focus; microreticulum.
8. Ornamentation in mesocolpium at low focus; columellae circular in outline, more or less inordinately arranged.

PLATE XIV (p. *121*)

Swertia perennis L. (Biol. Exc. France 1953-703)
1. Scanning electron micrograph; outline in polar view more or less angular, small scattered granules on colpus membrane (× 2000).
2. Scanning electron micrograph; ornamentation in mesocolpium; striato-reticulate pattern (× 5000).
3. Optical cross-section; outline in polar view.
4. Optical cross-section; outline in equatorial view.
5. Colpus membrane with small scattered granules.
6. Ornamentation in mesocolpium at high focus.
7. Ornamentation in mesocolpium at low focus; columellae irregular in outline and inordinately arranged.

PLATE XV (p. *122*)

Lomatogonium rotatum (L.) Fries ex Fernald (Hooker s.n.)
1. Scanning electron micrograph; equatorial view, rather narrow colpus, bridge over the endoaperture (× 2000).
2. Scanning electron micrograph; ornamentation in mesocolpium (× 5000).
3. Optical cross-section; outline in polar view, thick exine.
4. Ornamentation in apocolpium at low focus; thick columellae, inordinately arranged.
5. Endoaperture; broadly elliptic, truncate and with obtuse ends.
6. Ornamentation in mesocolpium at high focus; striato-reticulate pattern.
7. Ornamentation in mesocolpium at low focus; columellae inordinately arranged.

PLATE I

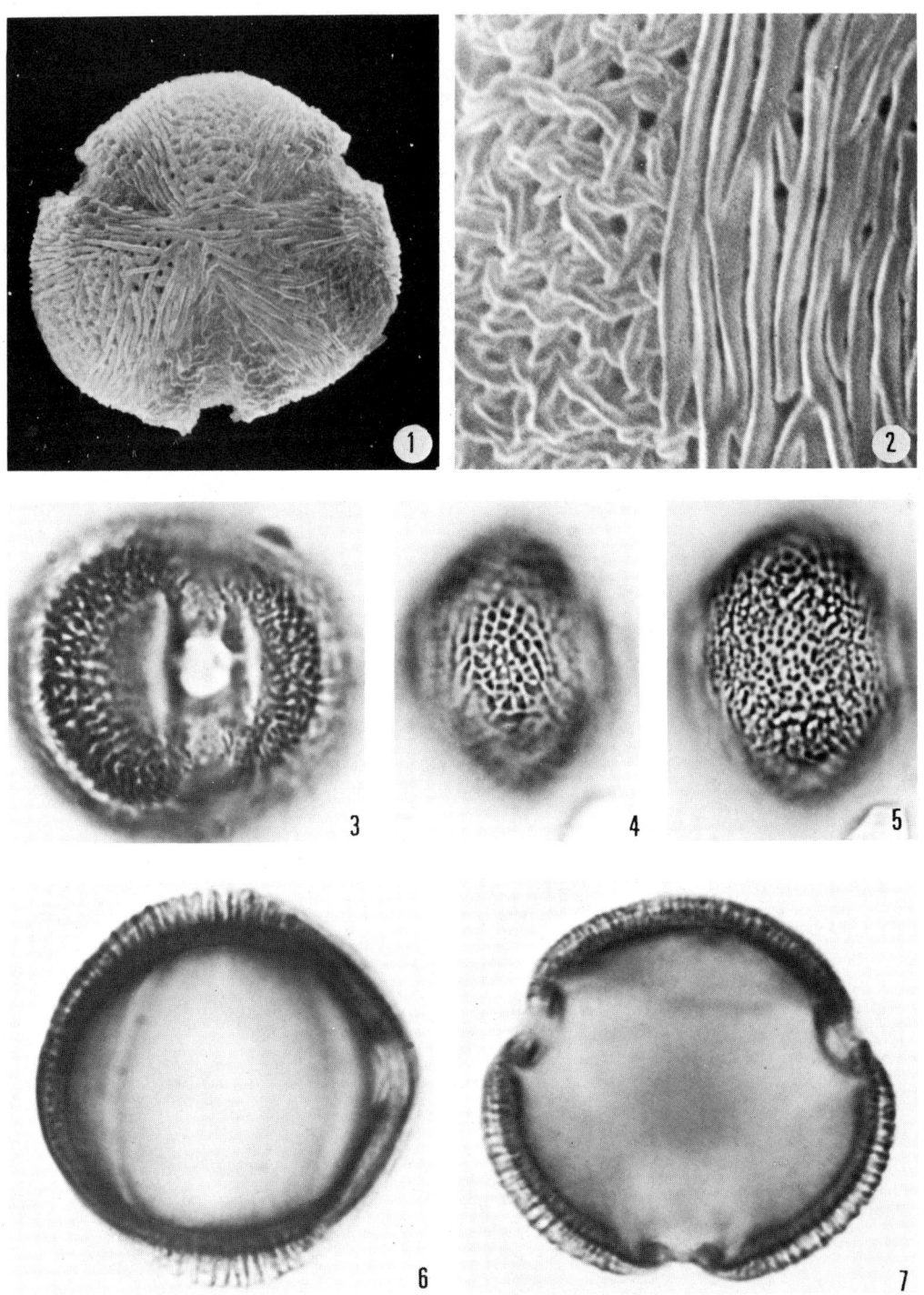

PLATE II

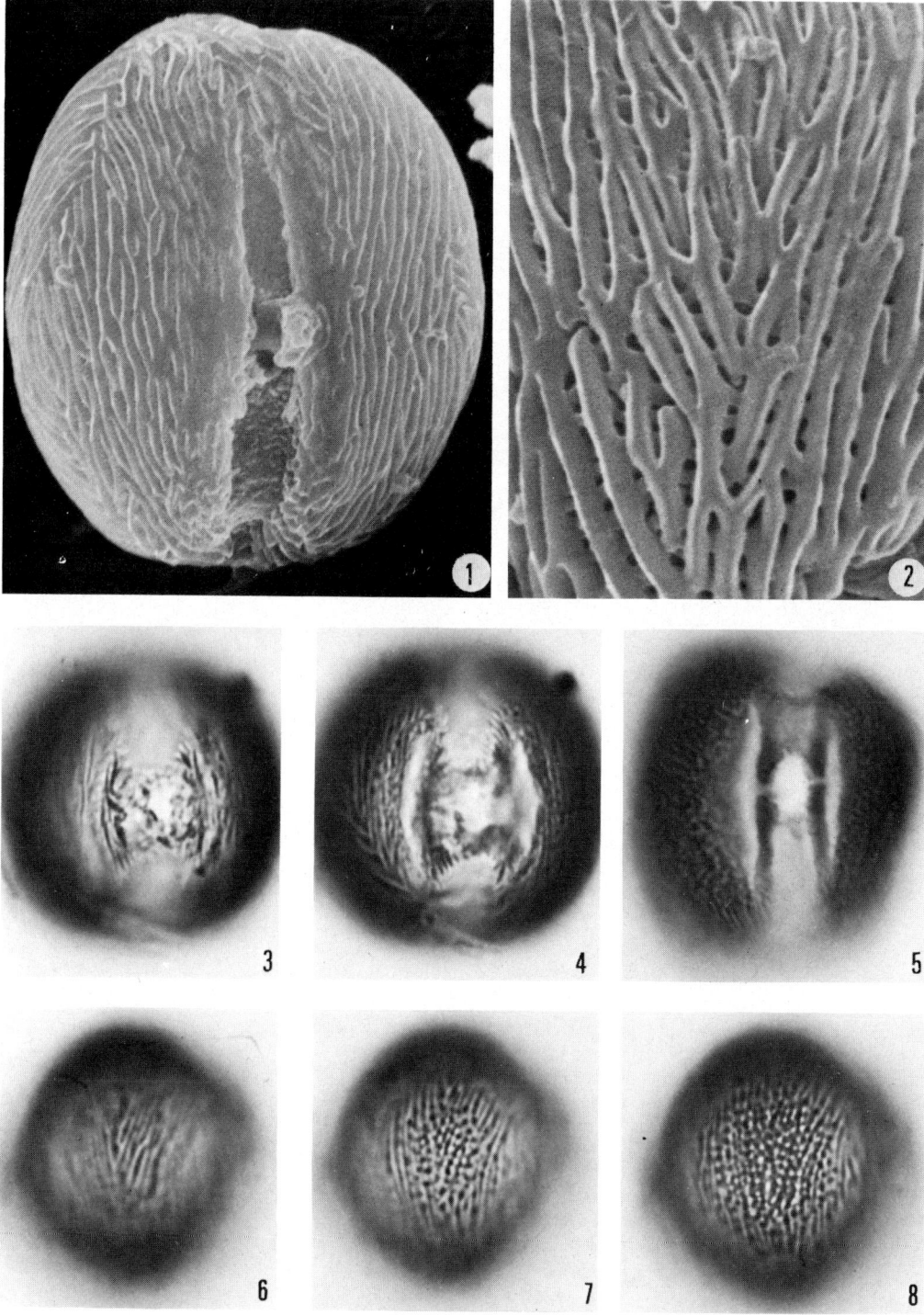

PLATE III

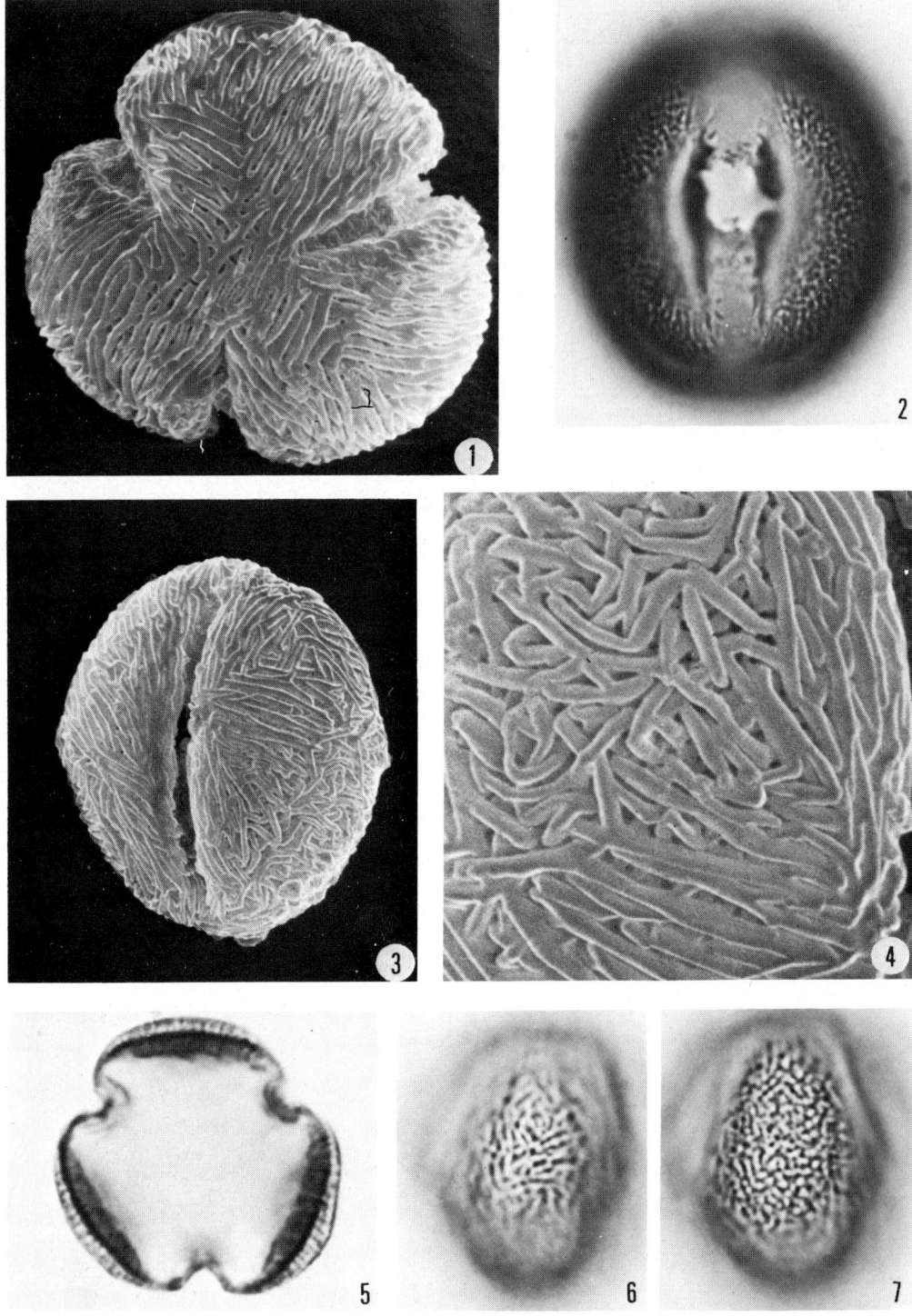

PLATE IV

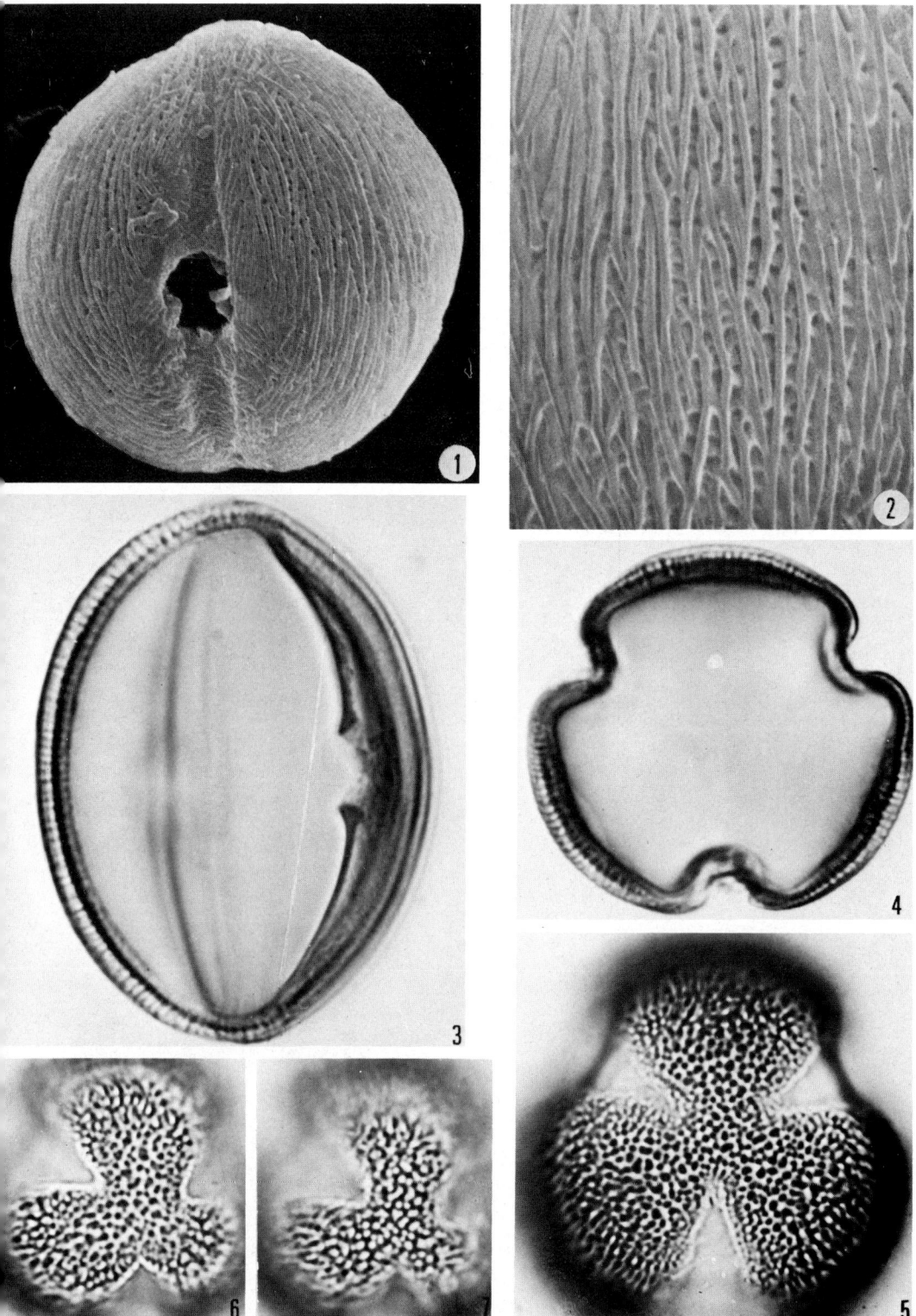

PLATE V

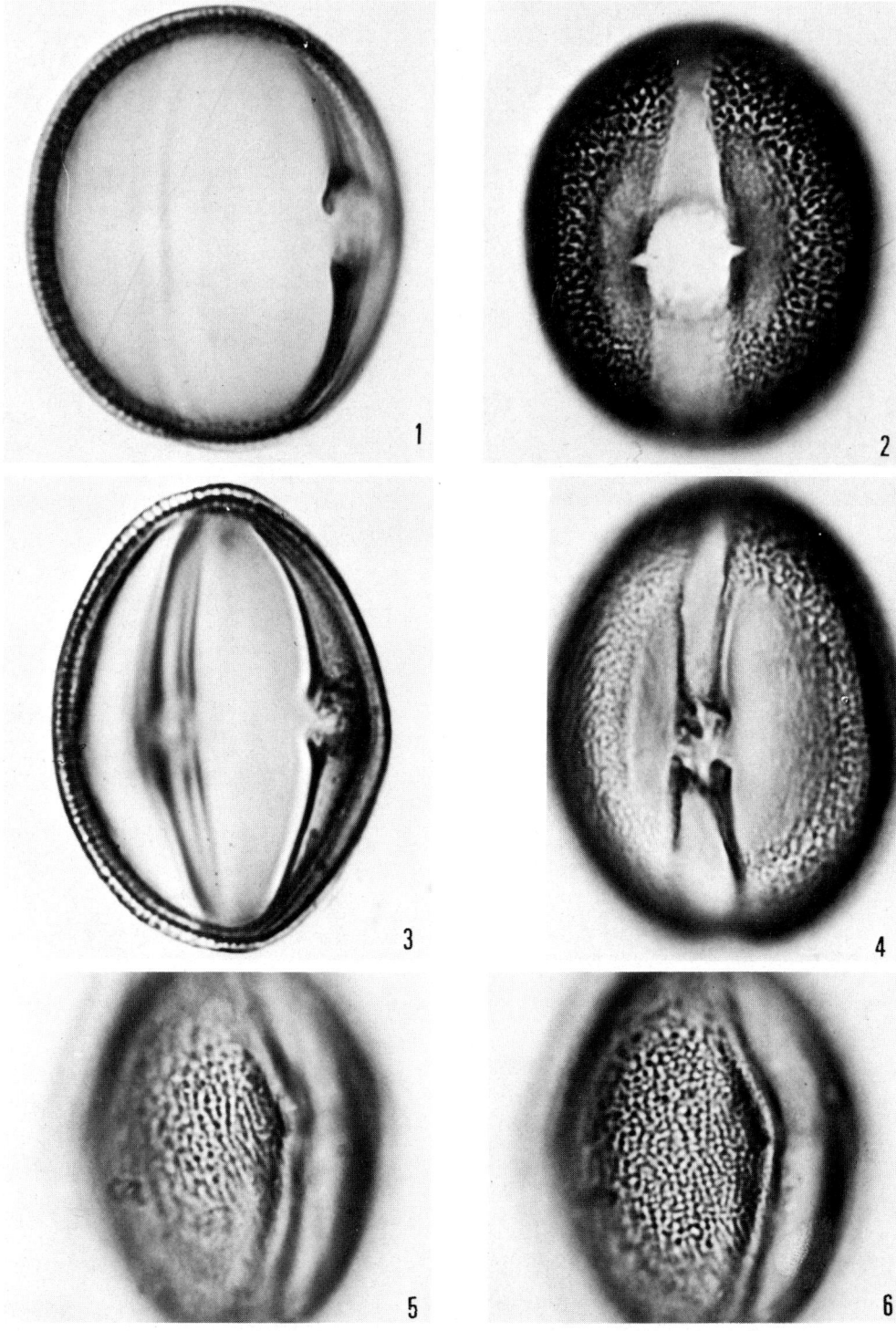

PLATE VI

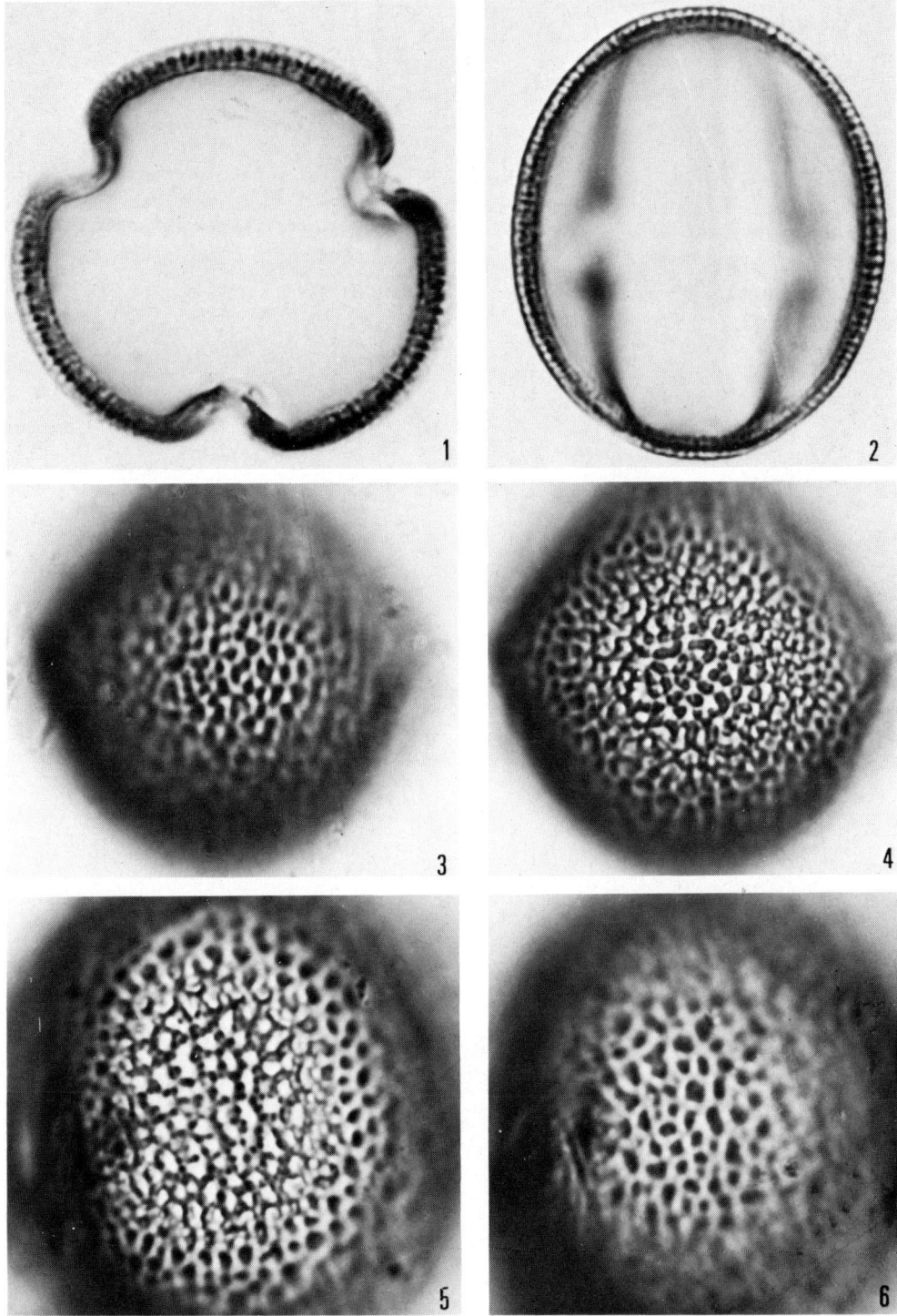

PLATE VII

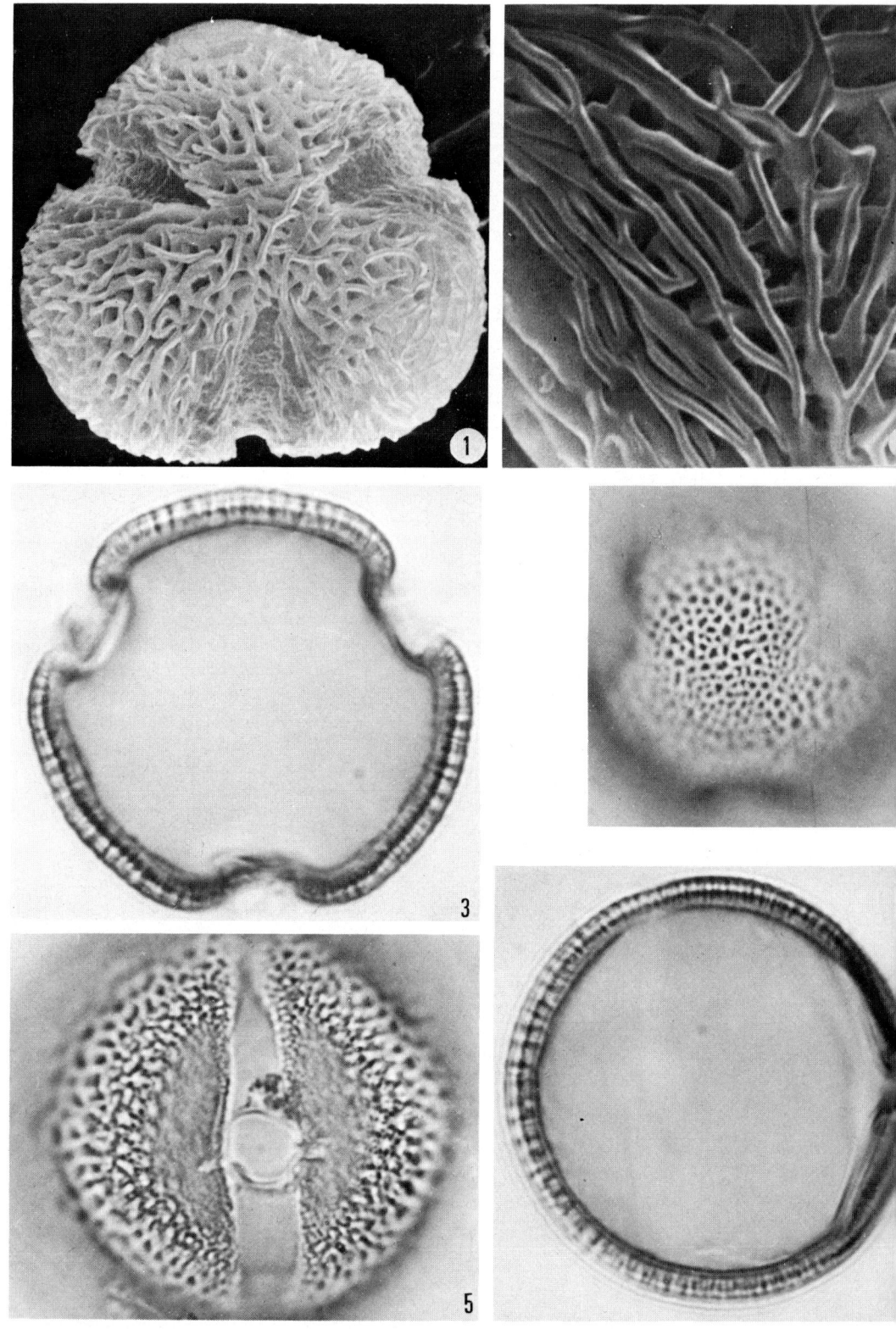

PLATE VIII

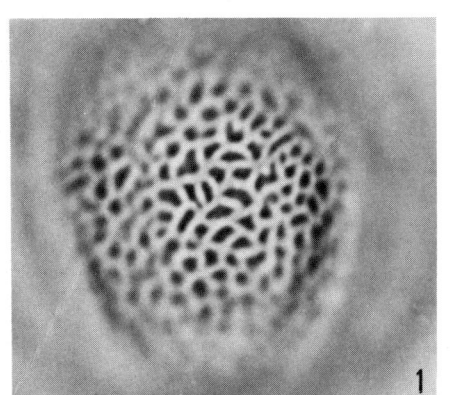
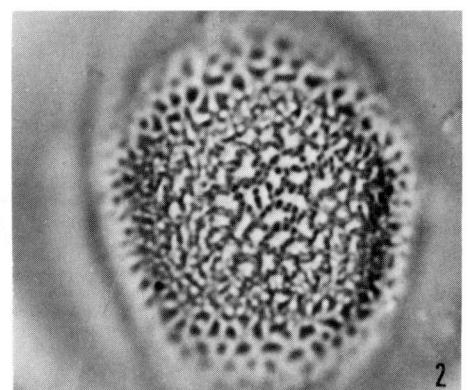
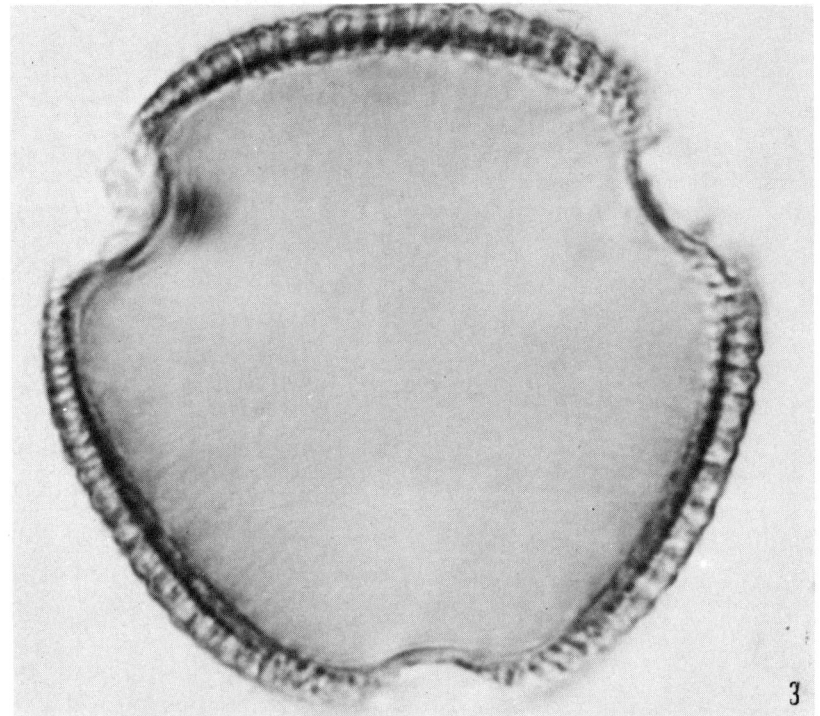
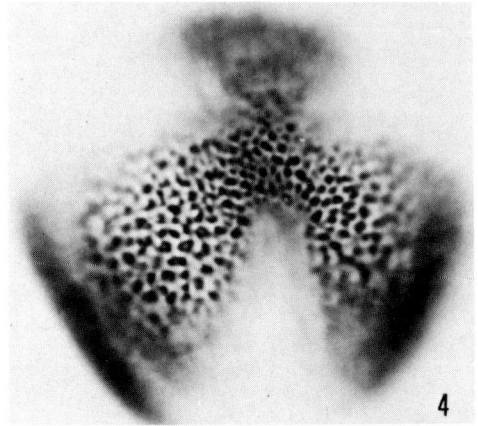
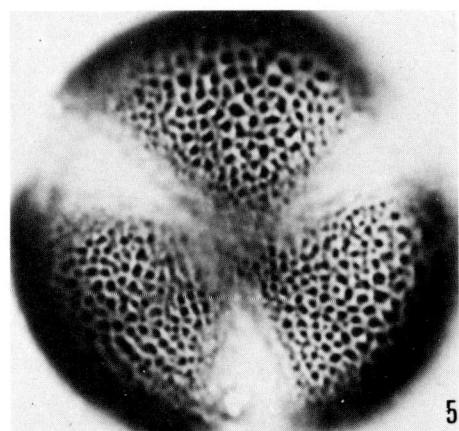

PLATE IX

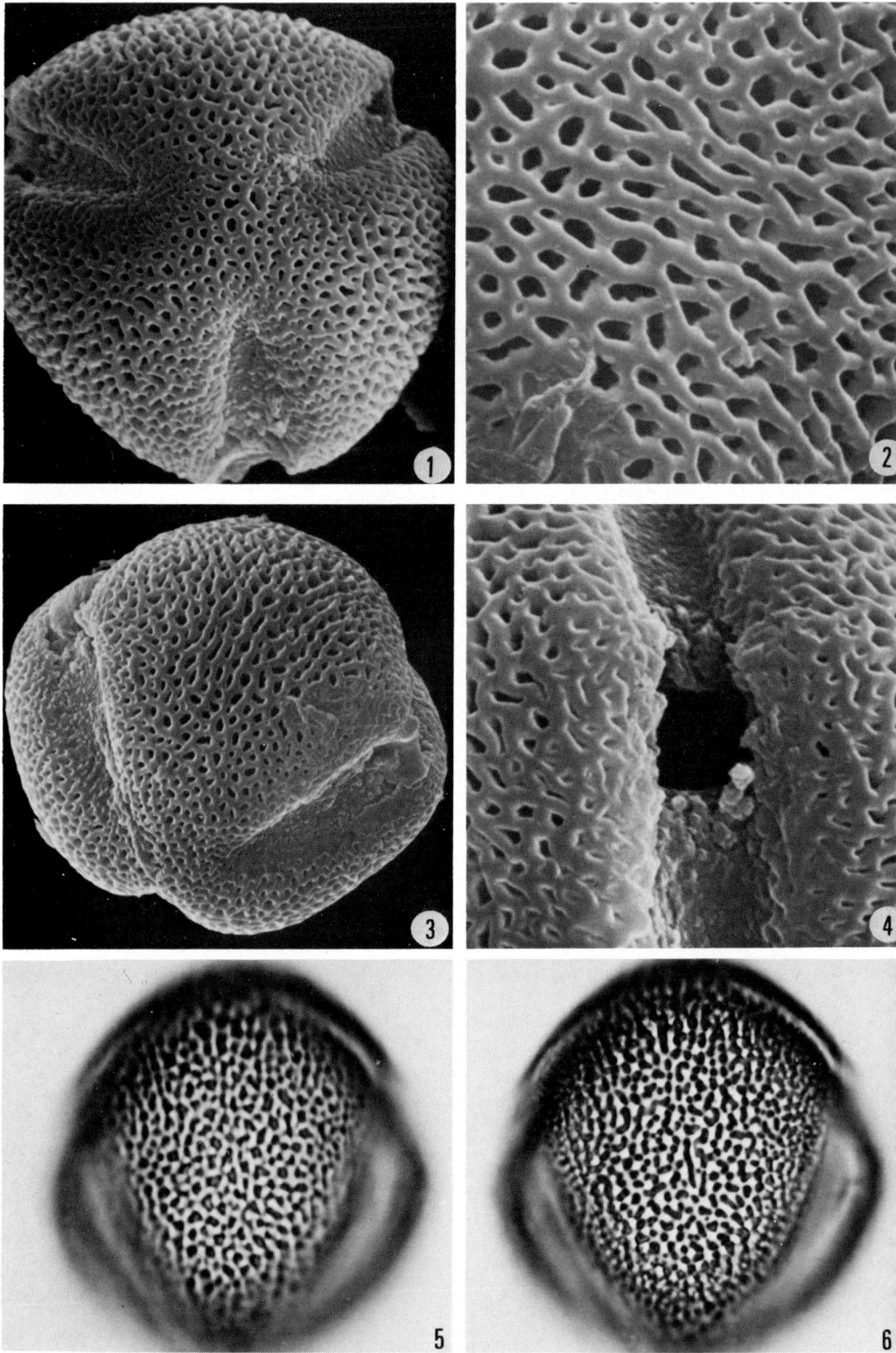

PLATE X

PLATE XI

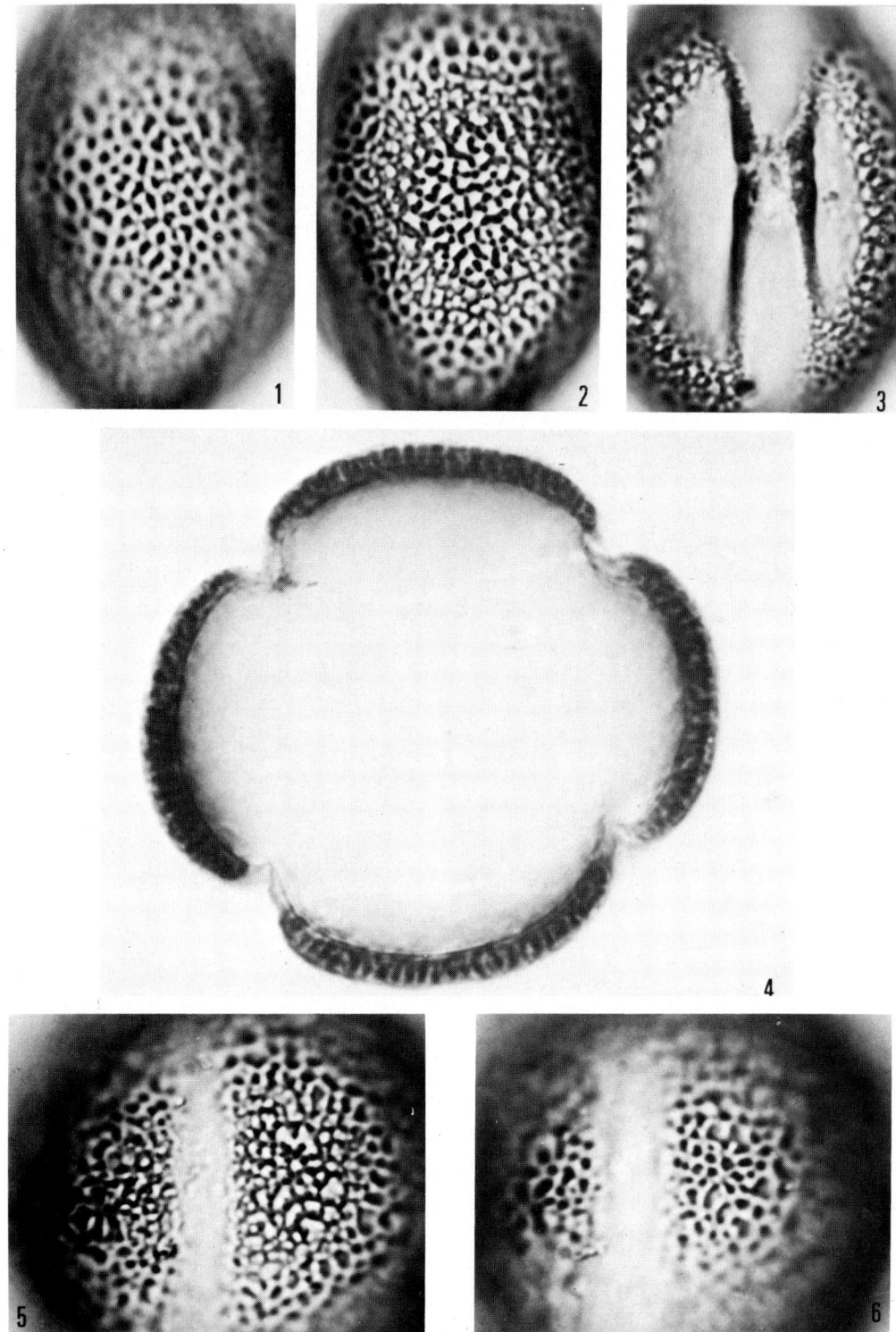

PLATE XII

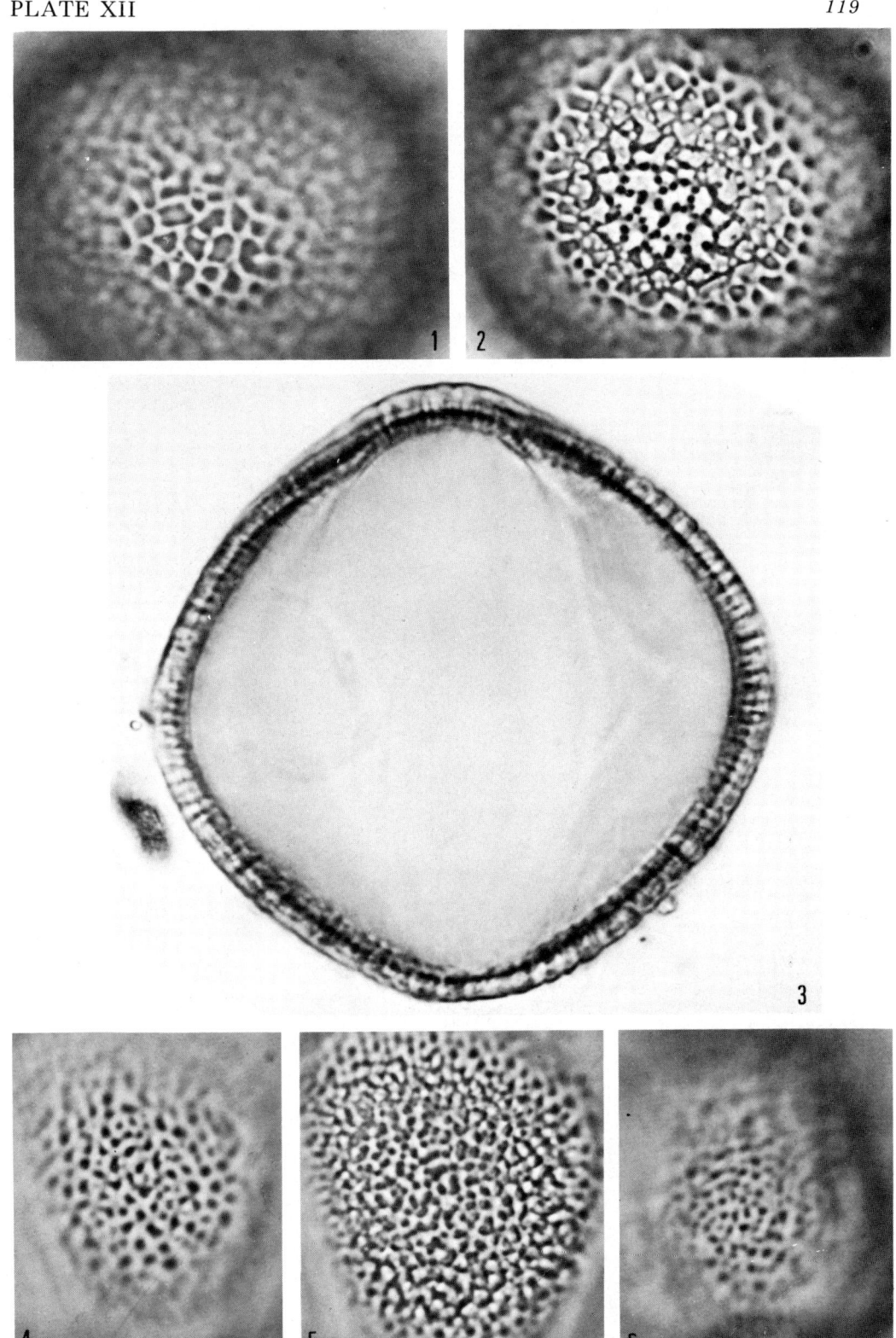

PLATE XIII

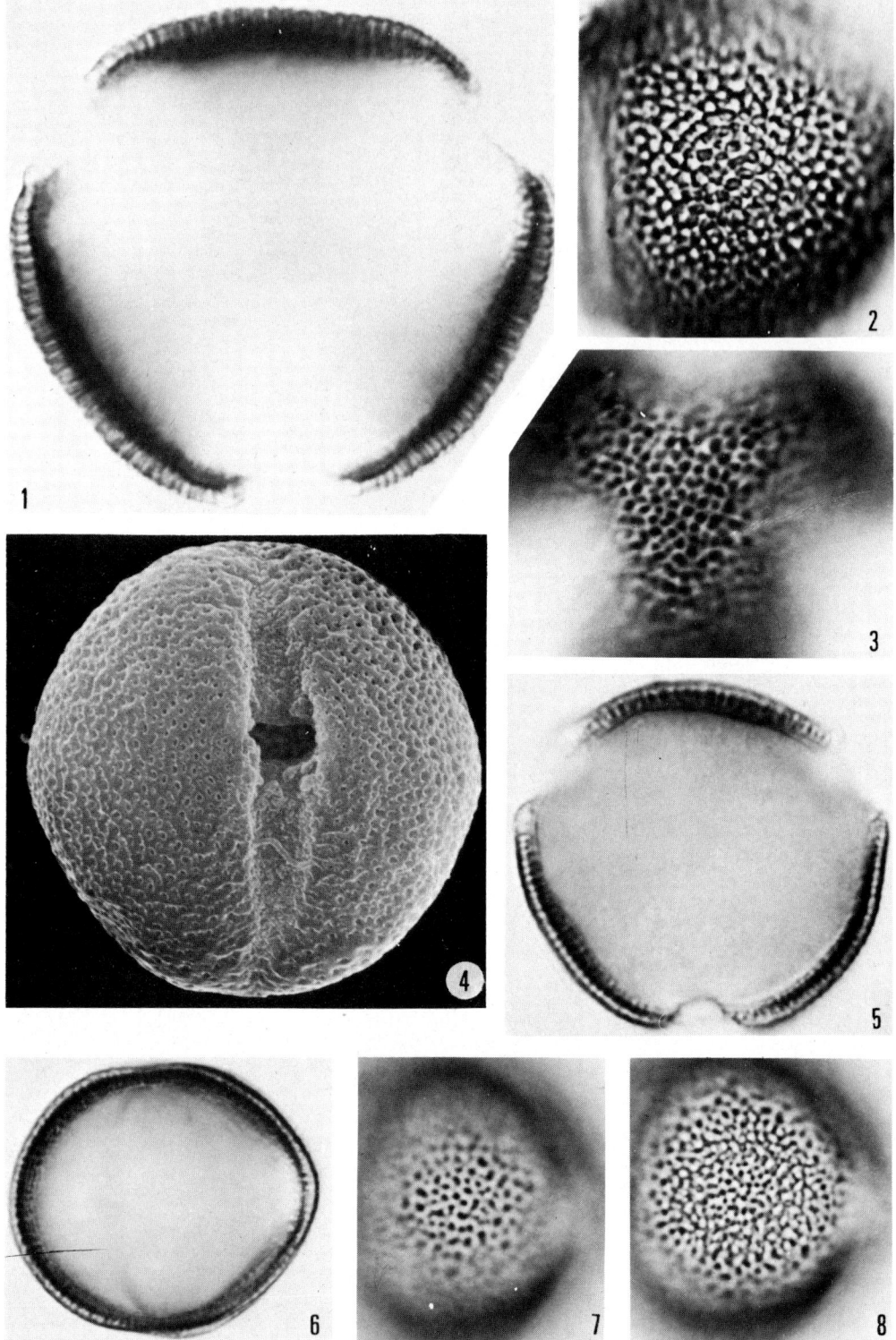

PLATE XIV

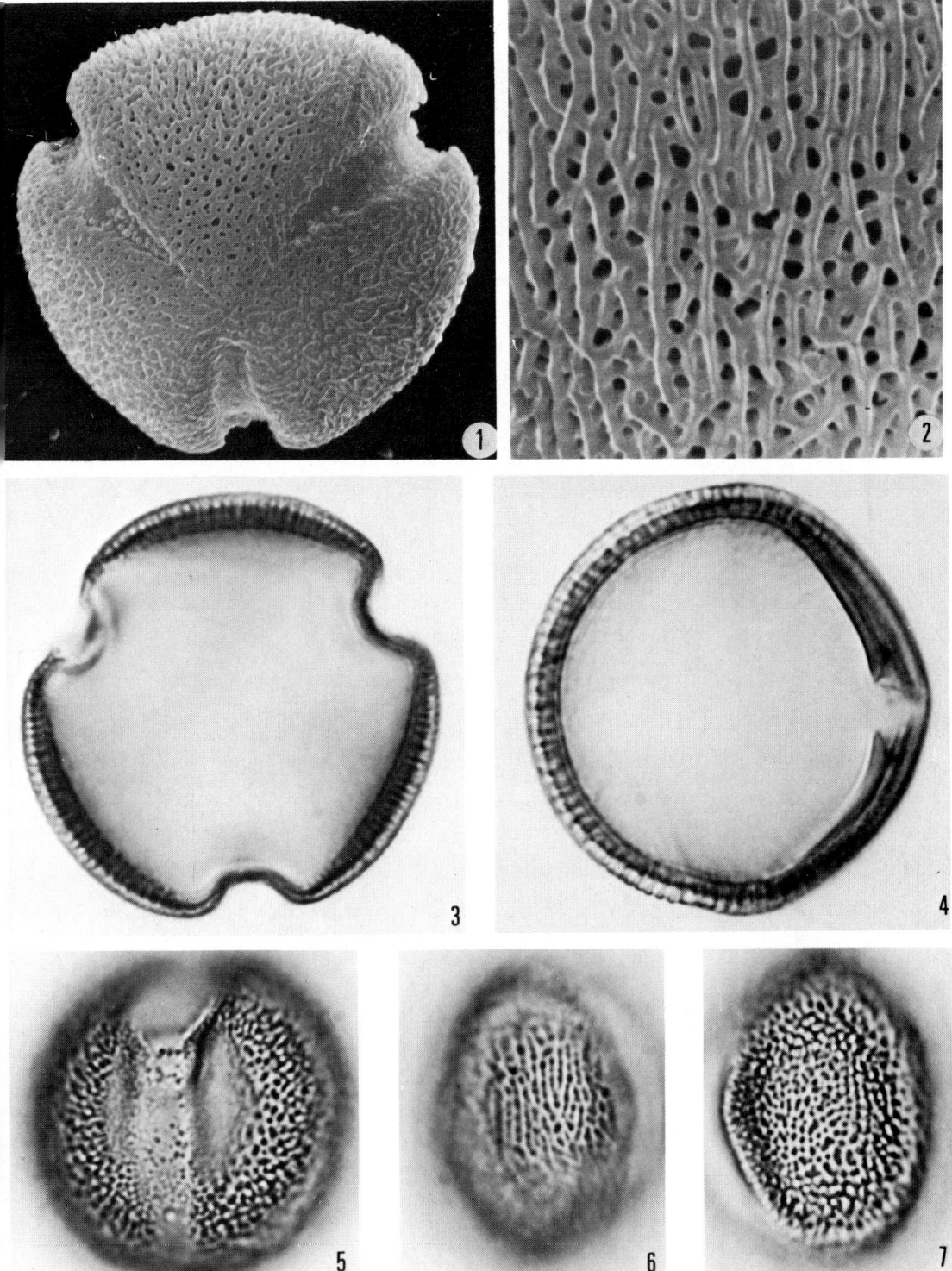

PLATE XV

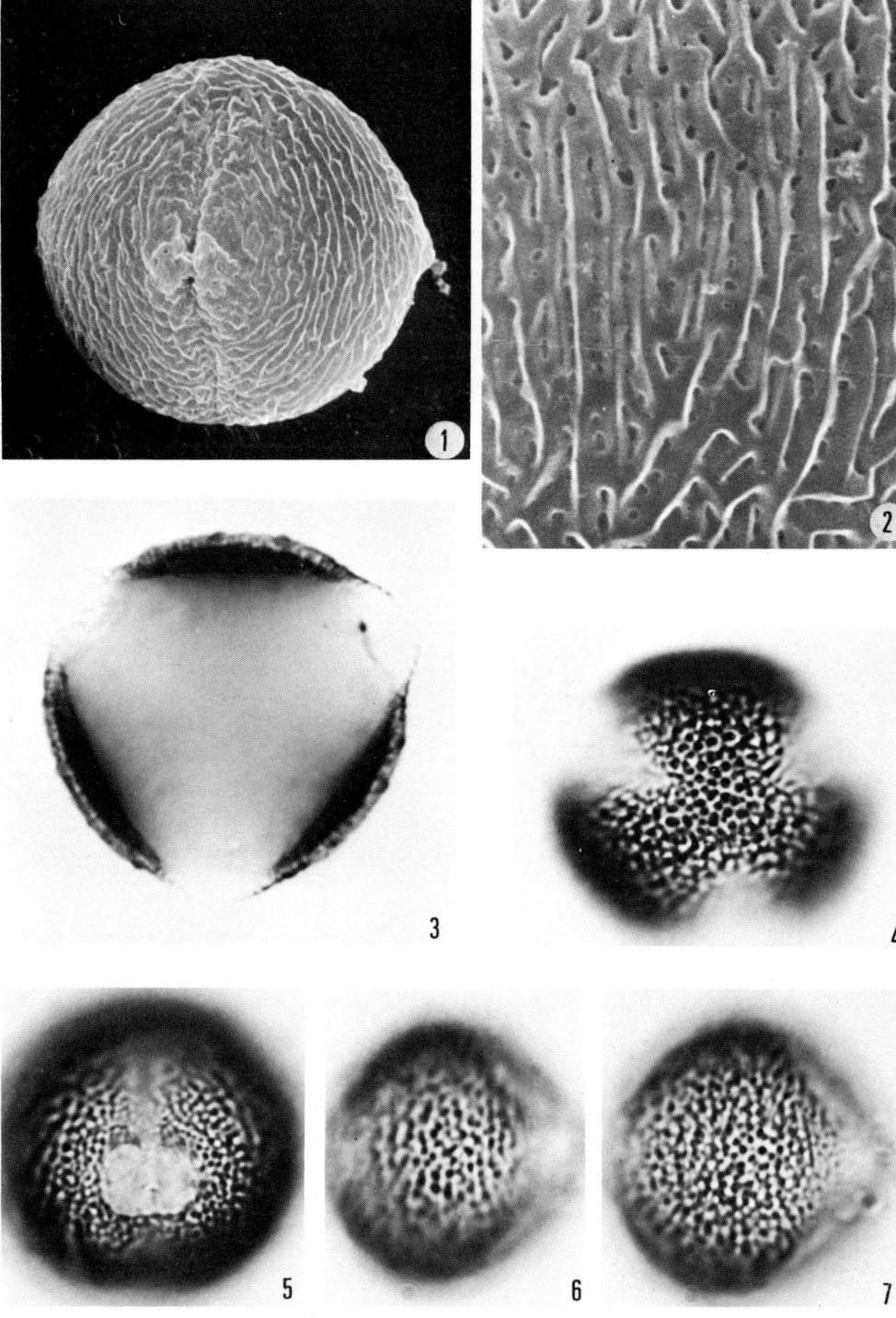

REFERENCES

Andersen, S. T., 1961. Vegetation and its environment in Denmark in the Early Weichselian Glacial (Last Glacial). Geol. Surv. Denmark, Ser. 2, 75 (Gentianaceae p.25).

Erdtman, G., 1943. An Introduction to Pollen Analysis. Chronica Botanica Co., Waltham, Mass., 239 pp.

Erdtman, G., 1952. Pollen Morphology and Plant Taxonomy. Angiosperms. Almqvist and Wiksell, Stockholm, 539 pp. (Gentianaceae p.183—185).

Erdtman, G., 1969. Handbook of Palynology. Munksgaard, Copenhagen, 486 pp.

Erdtman, G., Berglund, B. and Praglowski, J., 1961. An Introduction to a Scandinavian Pollen Flora, I. Almqvist and Wiksell, Stockholm, 92 pp.

Faegri, K. and Iversen, J., 1964. Textbook of Pollen Analysis. Munksgaard, Copenhagen, 2nd ed., 237 pp.

Gilg, E., 1897. Gentianaceae. In: Engler-Prantl, Die Natürlichen Pflanzenfamilien, IV, 1(2): 50—108.

Jonker, F. P., 1950. Revisie van de Nederlandse Gentianaceae I. *Centaurium* Mill. Nederl. Kruidkd. Arch., 57: 169—198.

Kuprianova, L. A. and Alyoshina, L. A., 1972. Pollen and Spores of Plants from the Flora of the European Part of the U.S.S.R. I. Acad. Sci. U.S.S.R., Leningrad, 171 pp.

Nilsson, S., 1964. On the pollen morphology in *Lomatogonium* A. Br. Grana Palynol., 5: 298—329.

Nilsson, S., 1967a. Pollen morphological studies in the Gentianaceae — Gentianinae. Grana Palynol., 7: 46—147.

Nilsson, S., 1967b. Notes on pollen morphological variation in Gentianaceae — Gentianinae. Pollen Spores, 9: 49—58.

Reitsma, Tj., 1970. Suggestions towards unification of descriptive terminology of angiosperm pollen grains. Rev. Palaeobot. Palynol., 10: 39—60.

Tarnavaschi, T. and Serbanescu-Jitariu, G., 1962. Recherches morphologiques concernant les microspores de quelques représentants de la famille Gentianaceae de la flore dans la R. P. Roumaine. Anal. Univ. C.I. Parhon, Ser. Stiint. Nat., 38: 119—131.

Tutin, T. G. and Melderis, A., 1972. Gentianaceae. In: T. G. Tutin, V. Heywood, N. A. Burges, D. N. Moore, D. M. Valentine, S. M. Walters and D. A. Webb (Editors), Flora Europaea. Cambridge, Univ. Press, pp.56—67.

The Northwest European Pollen Flora, 7

GUTTIFERAE

G. C. S. CLARKE

British Museum (Natural History), London (Great Britain)

LITERATURE

Aytuğ et al. (1971), Coutinho (1950), Erdtman (1952), Erdtman et al. (1961), Erdtman et al. (1963), Faegri and Iversen (1964), Fischer (1890), Kuprianova and Alyoshina (1972), Parmentier (1901).

INTRODUCTION

The family Guttiferae is represented in Europe by the single genus *Hypericum*. Previous publications mentioning the pollen of *Hypericum*, such as those of Erdtman et al. (1961) or Khan (1969) have commented on the similarity of the species. A current investigation of pollen morphology in the genus on a world scale is confirming this impression but showing that there are, nevertheless, a number of distinct pollen types which can be recognized. Most of the critical characters in these types are found in the nature of the exine sculpturing and the size and outline of the endoaperture. The former is most easily seen with the scanning electron microscope and the latter is often very hard to distinguish because of the construction of the ectoaperture, but several pollen types can be recognized using characters visible with the light microscope. Phase contrast can make exine-sculpturing features clearer but is not essential.

The taxonomy and selection of species in the area covered by this account follows that of Robson (1968).

SPECIMENS EXAMINED

All the specimens listed here are housed in the herbarium of the British Museum (Natural History) unless otherwise stated.

Hypericum androsaemum L. — England: Ladbrook s.n.; France: Fleming s.n.; Ireland: Foggitt s.n.; Portugal: Murray s.n.
H. calycinum L. — England: Thistleton-Dyer s.n.; Turkey: Rycroft s.n., Koe s.n.; Wales: Boswell Syme s.n.
H. canadense L. — Ireland: Webb s.n.; U.S.A.: Dress 35, Moldenke and Moldenke 9948.
H. elegans Stephan ex Willdenow — Germany: Rudolph s.n.; U.S.S.R.: Jakuscher 77.
H. elodes L. — Belgium: Bamps s.n., Devos, s.n.; England: Riddelsdell s.n.; France: Billot 2644, Heribaud s.n., Meinertzhagen s.n., Schultz 627.

H. gentianoides (L.) Britton, E. E. Sterns et Poggenburg — U.S.A.: Beattie s.n.
H. hircinum L. — Crete: Jermy and Brownsey 9031; England: White s.n.; France: Groves and Groves s.n.; Ireland: Carroll s.n.
H. hirsutum L. — Denmark: Flora Jutland. Exs. 705; England: Small s.n.; France: Atkins s.n.; Germany: Sonder s.n.; Sweden: Johanssen s.n.; Yugoslavia: Paulin 324.
H. humifusum L. — France: Deseglise s.n., Madiot 371; Germany: s.c., s.n.; Scotland: B. M. Mull Survey 4386; Spain: Elias s.n.; Sweden: Asplund s.n.
H. inodorum Miller — England: Evans s.n., Lousley s.n.
H. linarifolium Vahl — Channel Islands: Ward s.n.; France: Genevier s.n., Wilmott s.n.. Spain: Wilmott s.n.
H. maculatum Crantz ssp. *maculatum* — Belgium: Mosseray s.n.; Denmark: Young 4482; England: Dony, Dony and Allen s.n.; France: Lacaita 24977; Germany: Ham 1397; Norway: Williams s.n.
H. maculatum Crantz ssp. *obtusiusculum* (Tourlet) Hayek — Austria: Frohlich s.n.; Belgium: Alston s.n.; England: Atkins s.n., Dony s.n.; France: Williams s.n.
H. majus (A. Gray) Britton — Germany: Oberneder 3264; U.S.A.: Boufford 7592, Fernald 238.
H. montanum L. — Belgium: Martius s.n.; Denmark: Dahl s.n.; England: Watson s.n.; France: Genevier s.n.; Italy: Lacaita 23681; Sweden: Lindén s.n.
H. mutilum L. — Italy: Sevier s.n.; U.S.A.: Beattie s.n., De Wolf 1403.
H. perforatum L. — England: Jermy 226, Tittley 10; France: Carbonet s.n., Genevier s.n., Kendrick and Moyes 296; Germany: Holm-Nielsen et al. 16; The Netherlands: Rem s.n. (U); Sweden: Johanssen s.n.
H. pulchrum L. — Belgium: Louvabrée 11336; England: Gerrans 329; Faeroes: Orchard and Prince 35; France: Clement s.n.; Norway: Roikeland s.n.; Portugal: Fernandes and Matos 5579.
H. tetrapterum Fries — Belgium: Boulanger s.n., Lambert s.n.; Denmark: Holm-Nielsen and Jeppesen 347; England: Bowden and Hillman 259; France: Miciol s.n.; The Netherlands: Van Oort s.n. (U); Sweden: Sjörall s.n.
H. undulatum Schousboe ex Willdenow — England: Barton s.n.; France: Lloyd s.n.; Spain: Lacaita 17808.

KEY TO THE POLLEN TYPES

1.a. Pollen grains with tectum perforatum or microreticulate, lumina nowhere larger than 1 μm; columellae indistinct in L—O analysis. . . 2
 b. Pollen grains at least partly reticulate; columellae distinct in L—O analysis 3
2.a. Outline in polar view basically triangular with colpi inset in the angles of three slightly convex mesocolpia; endoaperture a distinct lalongate colpus . . . . *Hypericum calycinum* type
 b. Outline in polar view basically three-lobed with very convex mesocolpia separated by deep colpus grooves; endoaperture cruciform, sometimes very indistinct . . . . *Hypericum perforatum* type
3.a. Equatorial diameter 25 μm or more; endoaperture a circular or lalongate porus. . . . . *Hypericum elodes* type
 b. Equatorial diameter 20 μm or less; endoaperture a lolongate porus . . . . *Hypericum canadense* type

DESCRIPTION OF THE POLLEN TYPES

Hypericum calycinum type (Plate I; Plate II, 1—5)

Pollen class: 3-Zonocolporate (rarely 2- or 4-syncolporate).
P/E ratio: Suberect to erect.
Apertures: Ectoaperture — colpus, very long, occasionally syncolpate, narrowed or closed by raised sexine extensions at the equator; costae developed towards, but not above endoaperture, fastigium usually present, colpus membrane very faintly granulate under phase contrast. Endoaperture — colpus, lalongate, distinct although at least partly covered by sexine extensions, lateral ends rounded, obtusely pointed or jagged, costae absent.
Exine: More or less even thickness throughout but sometimes slightly thicker at poles. Nexine very weakly distinct from, and about equal in thickness to sexine. Sexine subdivisions indistinct except in phase contrast; columellae short.
Ornamentation: Tectum perforatum with superficial microreticulate pattern. SEM shows minute puncta at the base of conical depressions. Under light microscope lumina of microreticulate pattern irregularly rounded ca. 0.75 μm throughout or irregularly shaped lumina each containing several small perforations.
Outlines: Equatorial view — elliptic to slightly rhombic. Polar view — obtusely triangular with convex sides, angles truncated by colpi.
Measurements: Glycerine jelly — P 19—27 μm, E 14—23 μm (see table below for species details) exine ca. 1.5 μm; ectoaperture up to 3 μm wide; endoaperture 2—3 × 4—10 μm. Silicone oil — P 17—24 μm, E 13—21 μm, P/E ratio 1.2—1.7.

	P (μm)	E (μm)	Average P/E ratio
H. androsaemum	19—25	14—19	1.29
H. calycinum	23—27	17—23	1.23
H. hircinum	21—24	18—22	1.06
H. inodorum	21—24	17—20	1.15

Species: H. androsaemum, H. calycinum, H. hircinum, H. inodorum

Key to the species
N.B.: In this key the term "grouped perforations" has been used. This refers to an exine ornamentation composed of an ill-defined suprareticulum. The tectum underneath the irregularly shaped lumina are perforated by a series of puncta which consequently appear to be arranged in discrete groups, often sinuous lines, separated by the muri of the suprareticulum.
1.a. P/E ratio small (average value less than 1.20); endoaperture 7—10 μm wide, distinct 2
   b. P/E ratio large (average value more than 1.20); endoaperture 4—5 μm wide, often obscure 3

2.a. Inner wall of mesocolpial nexine always convex in polar view; ornamentation microreticulate or rarely with grouped perforations
. . . . H. hircinum
  b. Inner wall of mesocolpial nexine usually straight or concave in polar view; ornamentation often of grouped perforations. . . H. inodorum
3.a. Ornamentation of grouped perforations . . . . . H. androsaemum
  b. Ornamentation microreticulate . . . . H. calycinum

Comments

Taxonomically and palynologically this pollen type is well defined and although it is normally possible to distinguish the species included in it, one should not expect to key out every individual grain without some small proportion of misleading results. This is particularly true of specimens which have been in cultivation over many years or are growing outside their normal geographical and climatic range. In one of the specimens of *H. calycinum* examined in this survey, for example, (Wales, leg. Boswell Syme) about 25% of the pollen grains deviated from the usual tricolporate plan and were 2-syncolporate; in one specimen of *H. androsaemum* (England, leg. Ladbrook) the majority of the pollen grains were shrivelled and clearly infertile.

In the two Turkish specimens of *H. calycinum*, the average measurements of both polar axis and equatorial diameter were higher than those of the English specimens by some 9%. Too much should not be read into these figures as the number of samples taken into account is insufficient for an accurate analysis, but they suggest a possible phenomenon which should be borne in mind.

The most significant distinguishing character in this pollen type is the nature of the surface ornamentation, which varies from a regular superficial microreticulum in *H. calycinum* to a well-developed series of grouped perforations in *H. androsaemum*. This distinction is normally quite easy to see but it becomes considerably clearer if phase contrast is used.

The size and proportions of the endoaperture are also important. In *H. hircinum* and *H. inodorum* the equatorial width of the endocolpus is usually 7—10 μm while in *H. calycinum* and *H. androsaemum* it is rarely wider than 5 μm. This difference is paralleled by the extent to which the sexine covers the ectoaperture, and hence the clarity of the endocolpus, since a wider endoaperture is normally accompanied by a smaller sexine extension over the ectoaperture. On its own, the degree to which the ectoaperture is covered is, however, a poor diagnostic character since it varies considerably with the degree of expansion of the grains.

Hypericum canadense type (Plate II, 6—8; Plate III, 1—6)

Pollen class: 3-Zonocolporate (rarely 4-zonocolporate or with 6 or 8 apertures).
P/E ratio: Suberect to erect.
Apertures: Ectoaperture — colpus, long, more or less parallel-sided for most of its length or a little wider at equator, rarely slightly covered by sexine at

equator; slight costae present except above endoaperture; small fastigium present; colpus membrane smooth. Endoaperture — lolongate porus with clearly defined margins, occasionally with short, narrowed extensions on the meridional ends; costae faint.

Exine: Of equal thickness throughout grain. Nexine clearly distinguished from sexine, $\frac{1}{2}$—$\frac{2}{3}$ wall thickness. Sexine 1 columellate, thicker than semi-tectate sexine 2.

Ornamentation: Reticulate and microreticulate. Muri thin, slightly thicker towards base, simplicolumellate. Lumina irregularly angular in reticulum, more rounded in microreticulum; columellae also present in lumina (phase contrast), shorter than those in muri.

Outlines: Equatorial view — obtuse-rectangular to elliptic, the sides slightly convex, the poles rather flattened. Polar view — three convex lobes separated by deep colpi with U-shaped sinuses.

Measurements: Glycerine jelly — P 17—24 µm, E 12—17 µm (see table below for details of species); exine ca. 1.5 µm; ectoaperture up to 2.5 µm wide; endoaperture 5—8 × 4—6 µm. Silicone oil — P 15—22 µm, E 11—16 µm, P/E ratio 1.2—1.4.

	P (µm)	E (µm)	Average P/E ratio
H. canadense	17—23	13—17	1.29
H. majus	18—22	14—16	1.29
H. mutilum	17—24	13—16	1.36

Species: *H. canadense*, *H. majus*, *H. mutilum*

Key to the groups and species
1. a. Lumina of reticulum larger in mesocolpia than at poles . *H. canadense*
   b. Lumina of reticulum of similar size at poles and in mesocolpia or larger at poles . . . . *H. mutilum* group
   (*H. majus*, *H. mutilum*)

Comments

These species belong to the Section *Brathys* which is mainly distributed in North and South America (Keller, 1925). Robson (1968) cites all three as introductions to Europe, where *H. canadense* is the only one which is well naturalized. The other two species could hardly be said to be an established part of the flora of north-west Europe, but have been included because of their differing pollen morphology.

As a group the *H. canadense* type is very clearly distinct from all the others in the area, particularly in its exine ornamentation, which is reticulate rather than suprareticulate with small tectal perforations, but also in the lack of sexine extensions over the ectoaperture and in the lolongate rather than lalongate endoaperture. The variation in the way lumina size is graded across the exine surface in different species is a good distinguishing character. In all the species of this pollen type there are very small lumina (ca. 0.5 µm across) immediately next to the colpus and the size increases towards the mesocolpium. In *H. canadense* the lumina are very much larger in the mesocolpia than at the poles, while in the *H. mutilum* group the size of the lumina hardly varies between mesocolpia and poles, or has a tendency to be larger at the poles.

One specimen of *H. majus* from Germany (Oberneder 3264) shows a variety of pollen types. The standard tricolporate type is still the basic type, but there are also 4-colporate grains where the outline in equatorial view is rectangular and the colpi run diagonally across each face of the grain to join up and form a W-shape running round the grain. There are also very irregular grains with 6 or 8 apertures. All these pollen types have similar exine ornamentation but the distinction between lumen size in the mesocolpia and at the poles is less apparent with the increase of colpus number and consequent diminution of polarity.

H. gentianoides, which has been recorded as an introduction to southwest France (Robson, 1968) and also belongs to Section *Brathys*, has larger, more rhombic grains than any of the species in this pollen type and the lumina are smaller, more even in size but less regular in shape.

Hypericum elodes type (Plate III, 7, 8; Plate IV; Plate V, 1)

Pollen class: 3-Zonocolporate.
P/E ratio: Semi-erect to erect.
Apertures: Ectoaperture — colpus, very long, narrow at poles and widening gradually towards equator but partly covered at equator by broad sexine extensions, costae very well defined towards equator but terminated abruptly at sexine extensions; small fastigium occasionally present; colpus membrane very faintly granulate. Endoaperture — porus with short lateral extensions; costae present beneath ectoaperture but margins diffuse under sexine.
Exine: Of even thickness throughout mesocolpia, slightly thickened at poles. Nexine clearly distinguished from sexine varying from $\frac{1}{2}$ (at poles) to $\frac{3}{4}$ (in mesocolpia) of wall thickness. Sexine 1 of short, barely distinguishable columellae; sexine 2 semitectate.
Ornamentation: Suprareticulate with small perforations at the base of conical depressions. Muri thin at the top becoming wider below until they are $\frac{2}{3}$ the width of the lumina; columellae irregularly arranged with respect to muri which are usually duplicolumellate except at their junctions where they are clumped; under phase contrast shorter columellae are also visible in the lumina. Lumina irregularly rounded or slightly elongate, size variable, decreasing gradually towards poles and colpi.
Outlines: Equatorial view — elliptic tending to obtuse-rhombic. Polar view — obtusely triangular with rather convex sides, angles truncated by deep colpi.
Measurements: Glycerine jelly — P 35—45 μm, E 26—28 μm, P/E ratio 1.2—1.5; exine 2.5—3.0 μm; width of ectocolpus up to 3 μm; endoaperture 5—6 × 6—7 μm. Silicone oil — P 33—41 μm, E 24—34 μm, P/E ratio 1.3—1.5.

Species: *H. elodes*

Comments
H. elodes is a tetraploid species (Robson and Adams, 1968) and this is reflected in the increased size of the pollen compared with other related species. It is not only P and E which are large but also all the other features such as the reticulate ornamentation, wall thickness, columella size (and hence visibility), endoaperture dimensions, etc. It is as though the whole grain was magnified. This makes the pollen of *H. elodes* immediately recognizable and, although the increase of scale may not represent a fundamental difference from the pollen of such species as *H. montanum* (in the *H. perforatum* type), it seems well worthy of a pollen type of its own.

Hypericum perforatum type (Plate V, 2—5; Plate VI)

Pollen class: 3-Zonocolporate (rarely 4-zonocolporate or with 6 or 8 apertures).
P/E ratio: Semi-erect to erect.
Apertures: Ectoaperture — colpus, very long, more or less parallel-sided when seen from directly above but narrowed at equator by small, slightly raised sexine extensions; faint costae present except at poles and equator; fastigium present, but variable in extent. Endoaperture — cruciform porus or rarely colpus with meridional and equatorial extensions of varying lengths radiating from a centre, usually more or less obtusely angled, meridional ends rounded, costae sometimes present.
Exine: Of similar thickness throughout or sometimes slightly thicker at poles. Nexine weakly distinct from, and about as thick as, sexine. Sexine 1 of short columellae; sexine 2 semitectate.
Ornamentation: Tectum perforatum with superficial microreticulate pattern appearing faintly microreticulate in the light microscope. Muri about $\frac{1}{2}$ the width of lumina; columellae very small, only visible with phase contrast. Lumina rounded, more or less iso-diametric, of similar size throughout grain or slightly smaller near colpus.
Outlines: Equatorial view — elliptic with the poles sometimes a little pointed and the long sides occasionally almost straight. Polar view — three-lobed, deep colpus grooves alternating with convex mesocolpia; ectoaperture constrictions not prominent.
Measurements: Glycerine jelly — P 18—29 μm, E 11—19 μm (see table for details of species); exine ca. 1.0—1.5 μm; ectoaperture up to 2 μm wide; endoaperture 3—5 × 4—6 μm. Silicone oil — P 16—27 μm, E 10—17 μm, P/E ratio 1.4—1.55.

	P (μm)	E (μm)	Average P/E ratio
H. elegans	19—21	13—15	1.41
H. hirsutum	20—24	14—18	1.36
H. humifusum	18—25	12—16	1.46
H. linarifolium	21—26	14—17	1.44

	P (μm)	E (μM)	Average P/E ratio
H. maculatum			
ssp. *maculatum*	16—20	11—16	1.38
ssp. *obtusiusculum*	20—25	14—19	1.40
H. montanum	20—26	15—19	1.34
H. perforatum	21—29	13—19	1.45
H. pulchrum	22—26	14—21	1.41
H. tetrapterum	18—23	11—16	1.52
H. undulatum	18—23	12—16	1.45

Species: H. elegans, H. hirsutum, H. humifusum, H. linarifolium, H. maculatum, H. montanum, H. perforatum, H. pulchrum, H. tetrapterum, H. undulatum

Key to the groups and species
1.a. Lumina of suprareticulum normally between 0.75 and 1.0 μm across, endoaperture well-defined at least within ectocolpus 2
   b. Lumina of suprareticulum normally less than 0.75 μm across; endoaperture well defined or very obscure 3
2.a. Endoaperture clearly defined where it crosses ectocolpus and hardly less clear under sexine . . . . *H. montanum*
   b. Endoaperture much less clearly defined under sexine than in ectocolpus
      . . . . *H. pulchrum*
3.a. Endoaperture with meridional extensions which have slight costae, limits seen under sexine as rounded or obtusely pointed
      . . . . *H. perforatum* group
      (*H. elegans, H. hirsutum, H. maculatum, H. perforatum, H. tetrapterum, H. undulatum*)
   b. Endoaperture without thickened meridional extensions, almost invisible under sexine. . . . . *H. humifusum* group
      (*H. humifusum, H. linarifolium*)

Comments

Most of the common northwest European species of *Hypericum* fall into this pollen type and it is often very difficult to separate them. A key to a number of more or less recognizable groups is given but it would be wrong to suggest that every individual grain can be named successfully by using it. The range of variation of the species and the nature of the characters used is such that the key is better seen as a guide to identification of the groups rather than as providing an absolute means for their separation.

Of the species included in this type, *H. montanum* is the most easily recognized from its pollen and the most separate in its systematic position (Robson, 1968). It is more closely related to *H. elodes* than to the other species in the *H. perforatum* type and, apart from its size, its pollen is very similar to that of *H. elodes* except in the details of the endoaperture outline. Although *H. pulchrum* is similar in lumina size to *H. montanum*, the structure of its endoaperture is more like that of the *H. perforatum* group.

The outline of the endoaperture in the *H. humifusum* group is very hard to make out. It is not thickened where it crosses the ectocolpus and seems not to have any distinct meridional extensions (which are usual in the *H. perforatum* group). The endoaperture is very often completely covered by the extensions of the sexine even when the grains are expanded; this again is not usual in the *H. perforatum* group.

The *H. perforatum* group contains the species of Section *Hypericum* with the addition of *H. hirsutum*. The group is characterized by the combination of a small microreticulate pattern and a cruciform endoaperture which is relatively well defined where it crosses the ectocolpus but very obscure where it passes under the sexine of the ectocolpus margin. The meridional arms of the endoaperture often have slight costae at their tips and are apparently narrower than the equatorial arms which are not thickened.

No consistently reliable character was found for separating the species of this group.

The table of measurements shows that in *H. maculatum* the pollen of subspecies *obtusiusculum* is generally larger in both polar and equatorial diameter than that of ssp. *maculatum*. Robson (1957) has reported that while subspecies *maculatum* is diploid ($2n = 16$), subspecies *obtusiusculum*, the commoner subspecies in northwest Europe, is tetraploid ($2n = 32$), which no doubt explains the greater size of subspecies *obtusiusculum* pollen. The size difference between this diploid/tetraploid pair is however very much less than the difference between the tetraploid *H. elodes* and its diploid relatives, where the distinction in gross morphology is also much greater.

Many of the *Hypericum* species in northwest Europe, particularly those in the *H. perforatum* group, tend to produce occasional abnormal pollen types where the grains, although conforming to a recognizable series of morphological patterns, are often asymmetric and irregular in appearance. This phenomenon was originally commented on by Parmentier (1901) in *H. perforatum*. Staining tests on a number of herbarium specimens using the method of Alexander (1969) suggest that the abnormal grains are largely infertile. The simplest variant form is the 4-colporate grain with diagonally slanting colpi described under the *H. canadense* type (p. *128*). The commonest abnormal type, however, is a grain with six apertures where three colpi with endoapertures meet at one pole and are linked together at their other ends by three more colpi running across the polar axis. These three subsidiary colpi have no endoapertures. There is also an eight-apertured version of the same scheme. All these pollen types may be mixed in varying proportions with normal tricolporate grains in a single anther.

Noack (1939) reported that meiosis in *H. perforatum* is sometimes irregular and that this was a cause of pollen sterility. The abnormal pollen grains described here could well be correlated with such irregularities. Noack also found that *H. perforatum* was largely apomictic in its reproduction. This may be the reason why the normal tricolporate pollen grains in this species are more than usually variable.

PLATE I (For Plate descriptions see pp. *140–141*)

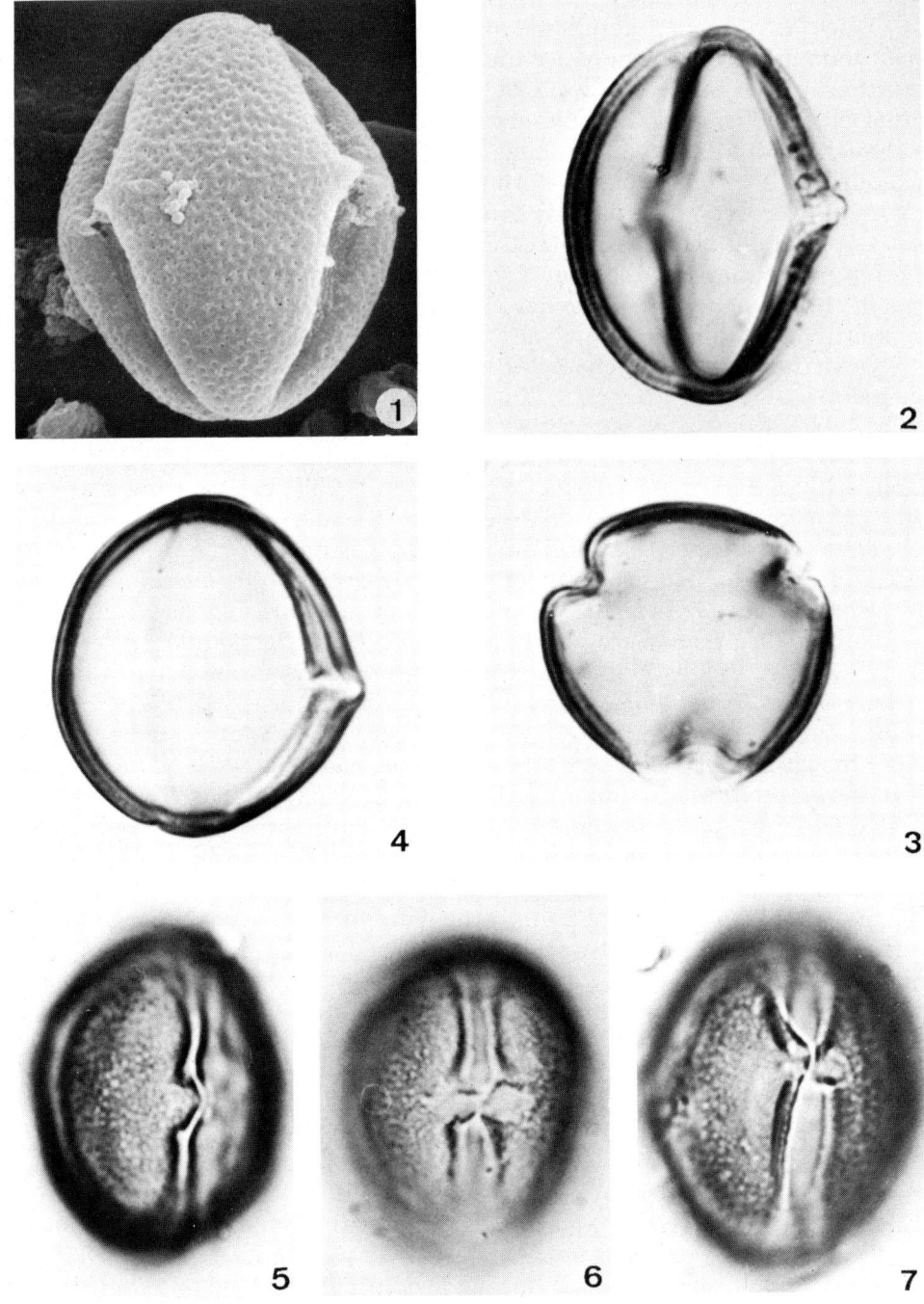

PLATE II

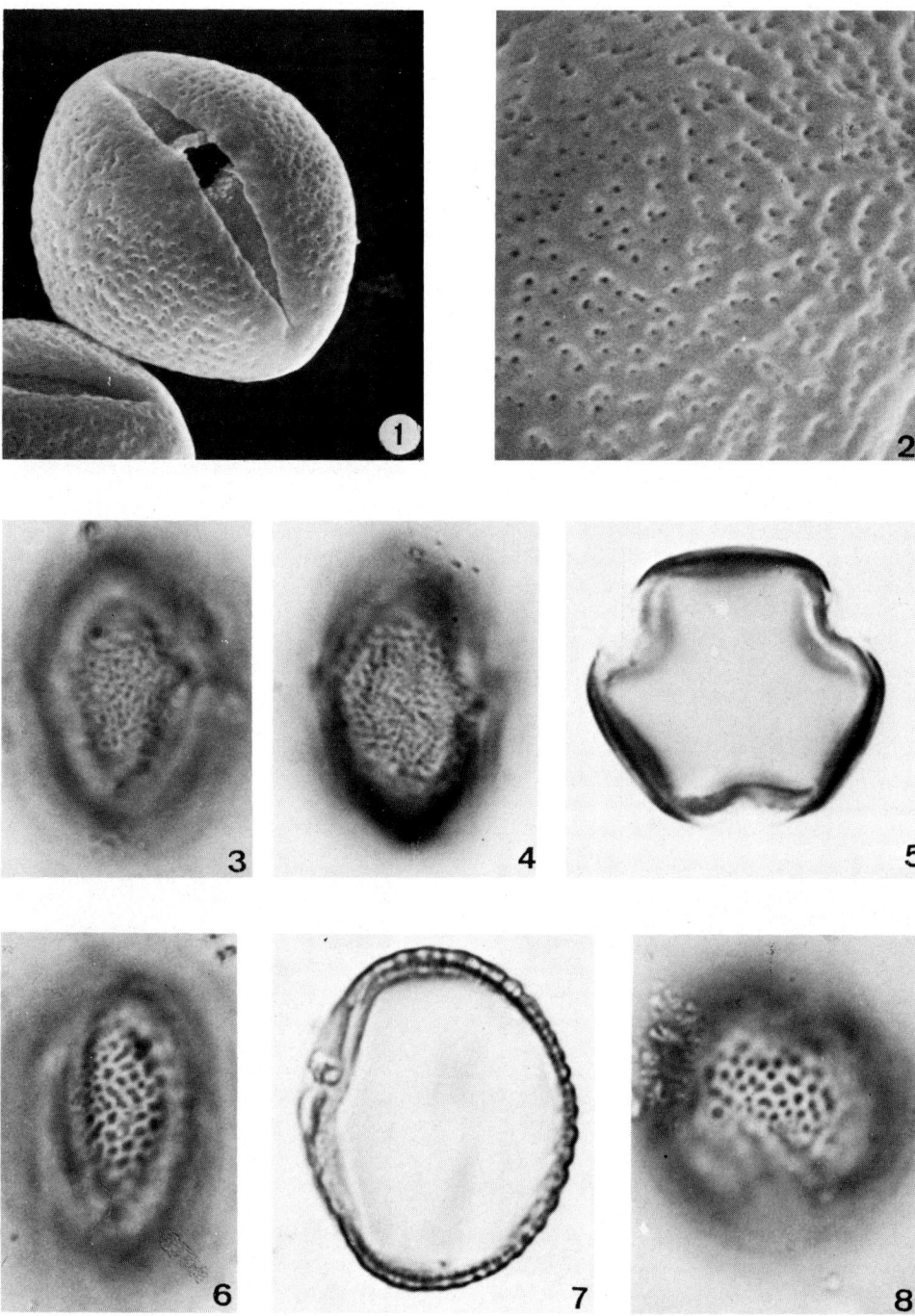

PLATE III

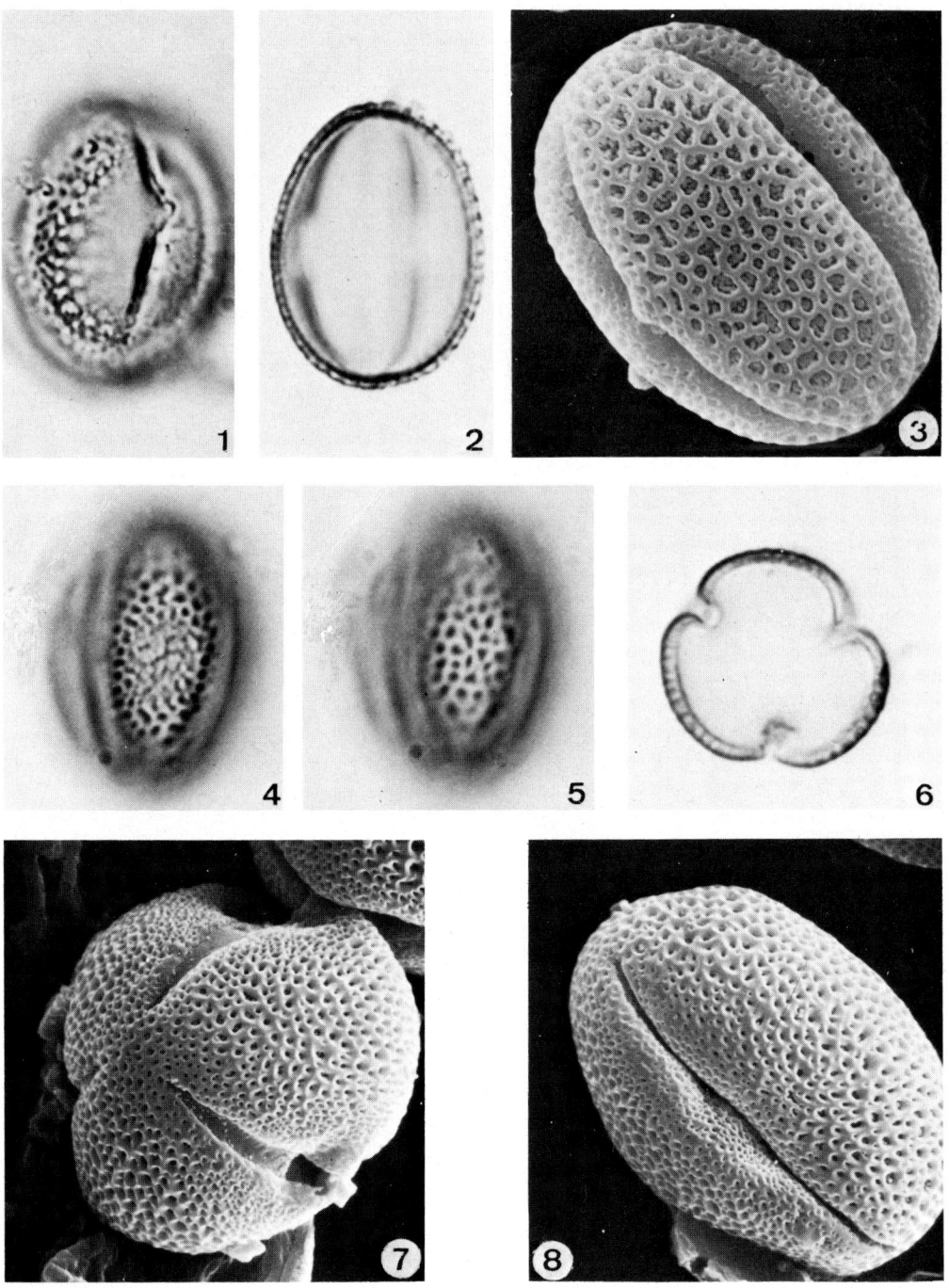

PLATE IV

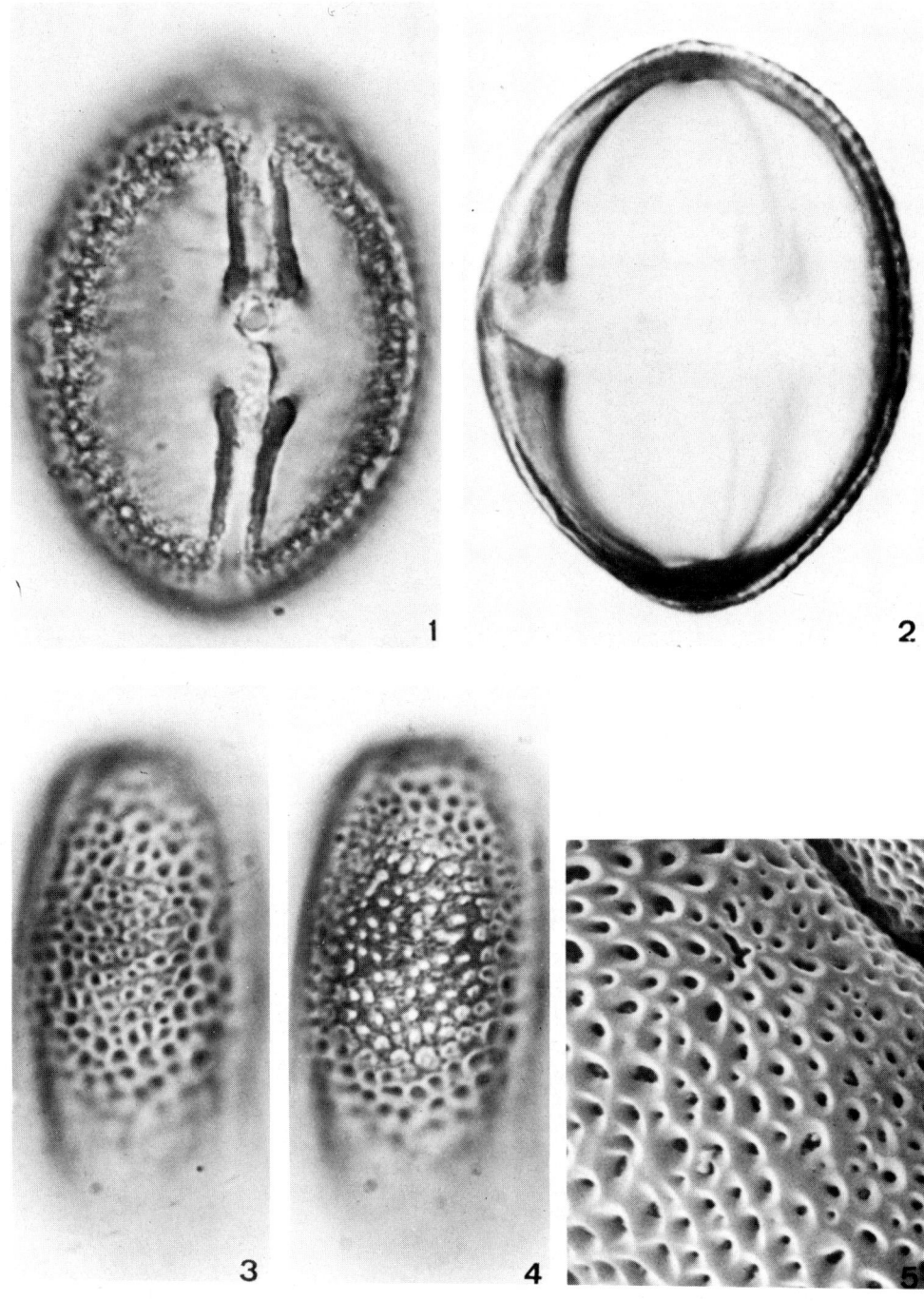

PLATE V

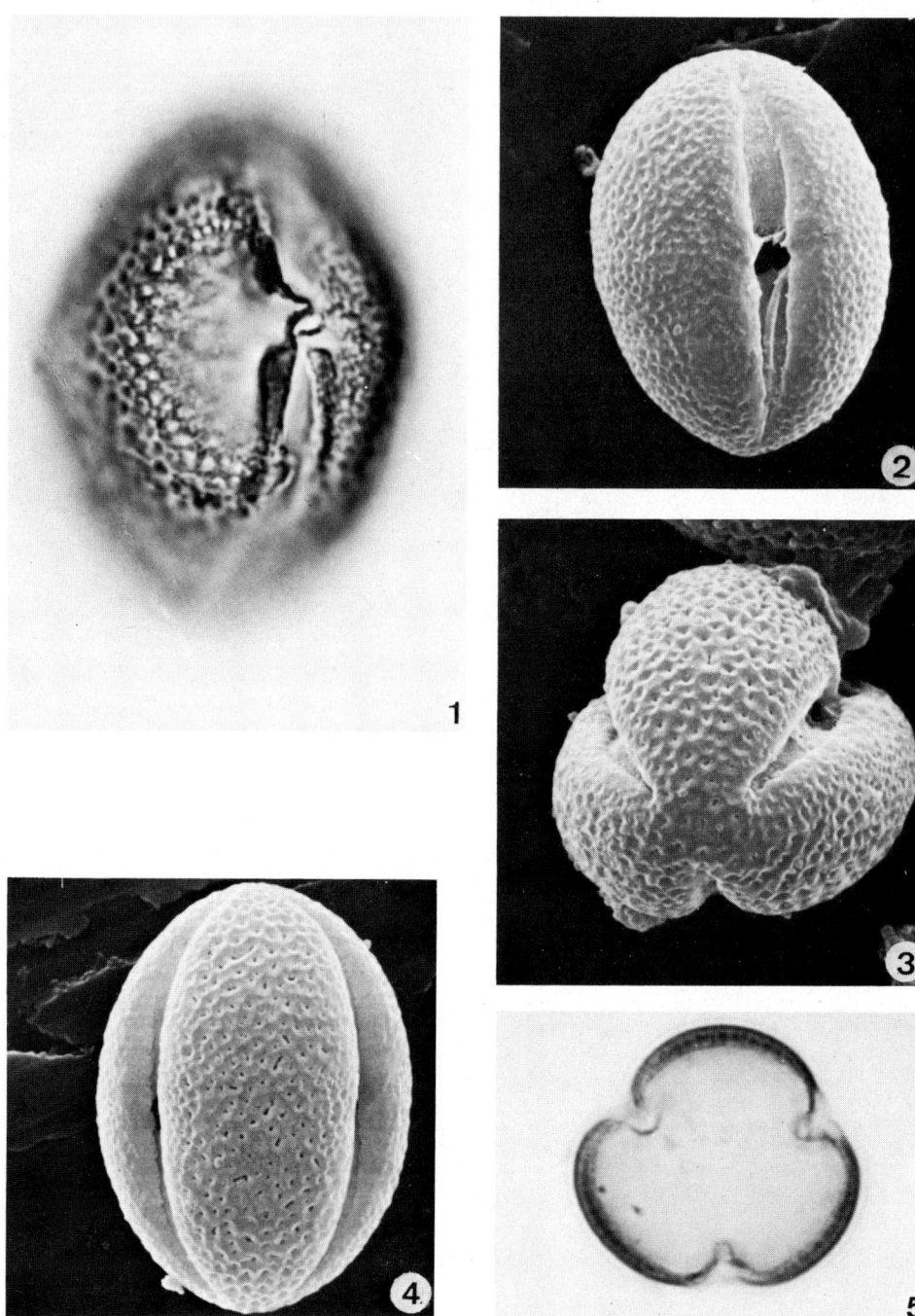

PLATE VI

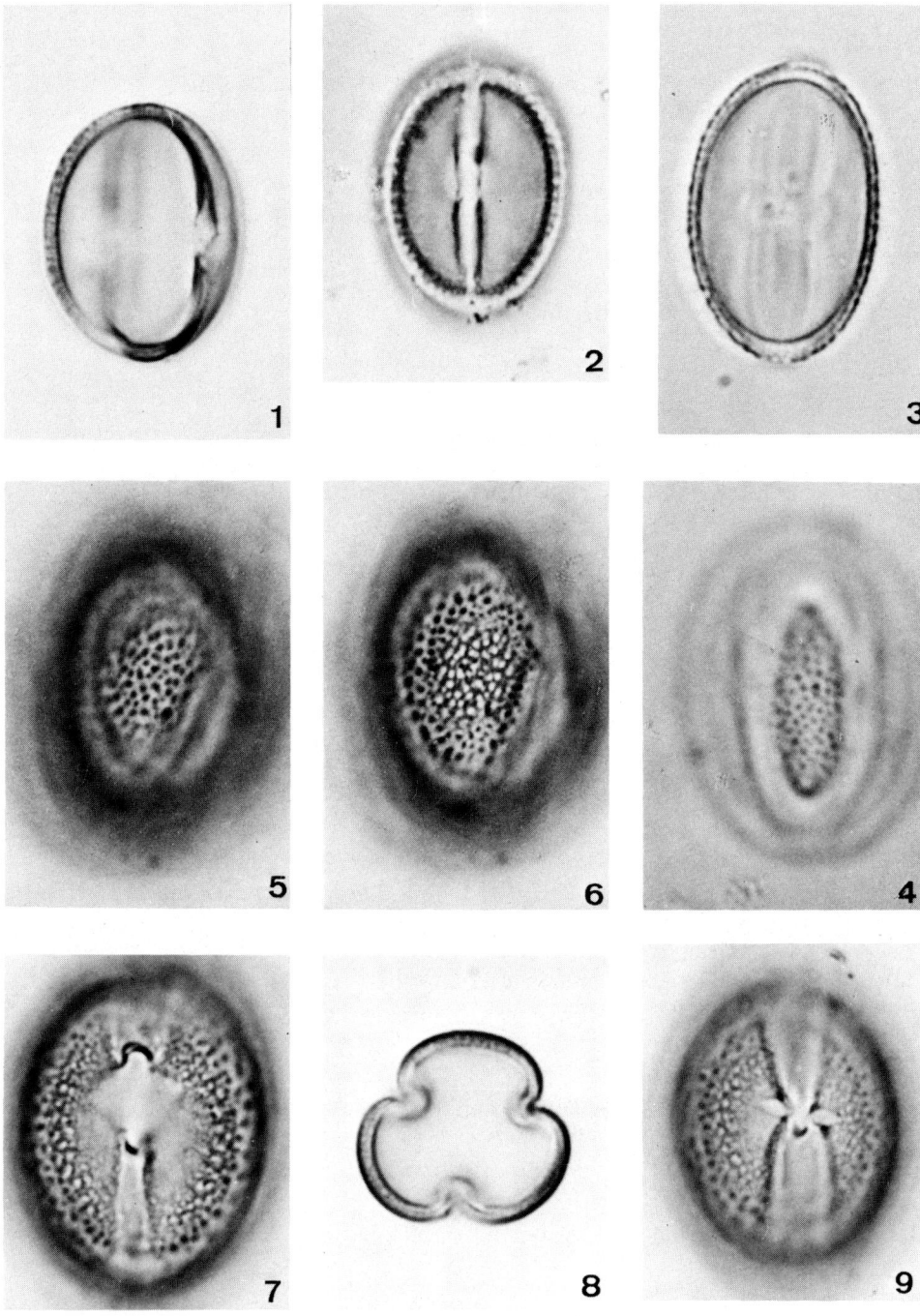

PLATE DESCRIPTIONS (all Plates × 2000, except as otherwise stated)

PLATE I (p. *134*)

Hypericum calycinum L. (fig.1, Rycroft s.n.; fig.2, 3, Thistleton-Dyer s.n.)
1. Scanning electron micrograph; overall shape (× 2500).
2. Equatorial view; cross-section, P/E ratio large.
3. Polar view; cross-section.
Hypericum hircinum L. (Groves and Groves s.n.)
4. Equatorial view; cross-section, P/E ratio small.
5. Equatorial view; fastigium.
6. Endoaperture, covered by sexine extensions.
Hypericum calycinum L. (Thistleton-Dyer s.n.)
7. Colpus, endoaperture and fastigium.

PLATE II (p. *135*)

Hypericum inodorum Miller (fig.1, 2, Lousley s.n.; fig.3—5, Evans s.n.)
1. Scanning electron micrograph; overall shape in equatorial view (× 2510).
2. Scanning electron micrograph; ornamentation, grouped perforations (× 6600).
3. Ornamentation at high focus.
4. Ornamentation at low focus.
5. Polar view; cross-section, inner wall straight to concave.
Hypericum canadense L. (Webb s.n.)
6. Ornamentation at mesocolpium at high focus; lumina large and angular.
7. Equatorial view; cross-section.
8. Ornamentation at apocolpium at high focus; lumina more or less rounded in outline.

PLATE III (p. *136*)

Hypericum canadense L. (Webb s.n.)
1. Equatorial view; colpus and fastigium.
Hypericum mutilum L. (Beattie s.n.)
2. Equatorial view; cross-section.
3. Scanning electron micrograph; overall shape in equatorial view (× 3300).
4. Ornamentation at low focus.
5. Ornamentation at high focus.
6. Polar view; cross-section.
Hypericum elodes L. (Billot 2644)
7. Scanning electron micrograph; polar view (× 1720).
8. Scanning electron micrograph; equatorial view and colpus (× 1720).

PLATE IV (p. *137*)

Hypericum elodes L. (fig.1—4, Riddlesdell s.n.; fig.5, Billot 2644)
1. Equatorial view; colpus and endoaperture.
2. Equatorial view; cross-section.
3. Ornamentation at high focus.
4. Ornamentation at low focus.
5. Scanning electron micrograph; ornamentation (× 4290).

PLATE V (p. *138*)

Hypericum elodes L. (Riddlesdell s.n.)
1. Equatorial view; colpus with fastigium.

Hypericum perforatum L. (Tittley 10)
2. Scanning electron micrograph; overall shape in equatorial view (× 2260).
3. Scanning electron micrograph; polar view (× 3100).
Hypericum montanum L. (Dahl s.n.)
4. Scanning electron micrograph; overall shape in equatorial view (× 2760).
5. Polar view; cross-section.

PLATE VI (p. *139*)

Hypericum maculatum Crantz (Dony, Dony and Allen s.n.)
1. Equatorial view; cross-section, fastigium.
2. Equatorial view; cross-section, colpus with circular endoaperture.
Hypericum tetrapterum Fries (v. Oort s.n.)
3. Equatorial view; cross-section.
4. Ornamentation at high focus.
Hypericum montanum L. (Dahl s.n.)
5. Ornamentation at high focus.
6. Ornamentation at low focus.
7. Equatorial view; colpus with endoaperture.
Hypericum maculatum Crantz (Dony, Dony and Allen s.n.)
8. Polar view; cross-section.
Hypericum pulchrum L. (Clement s.n.)
9. Equatorial view; colpus with elongated, colpus-like endoaperture.

ACKNOWLEDGEMENTS

It is a pleasure to thank Dr N. K. B. Robson and Dr W. Punt for their helpful comments during the course of this study, and Miss Marilyn Jones for her technical assistance.

REFERENCES

Alexander, M. P., 1969. Differential staining of aborted and non-aborted pollen. Stain Technol., 44(3): 117—122.
Aytuğ, E., Aykut, S., Merev, N. and Edis, G., 1971. Atlas des Pollens des Environs d'Istanbul. Kutulmuş Matbaasi, Istanbul, 330 pp.
Coutinho, M. C. P., 1950. Notas sobre a constituição histó-anatómica das diversas espécies do genéro *Hypericum* L. existentes na serra do Gerês. Agron. Lusitana, 12(4): 517—549.
Erdtman, G., 1952. Pollen Morphology and Plant Taxonomy. An Introduction to Palynology, 1. Angiosperms. Chronica Botanica Co., Waltham, Mass.; Almqvist and Wiksell, Stockholm, 539 pp.
Erdtman, G., Berglund, B. and Praglowski, J., 1961. An introduction to a Scandinavian pollen flora. Grana Palynol., 2(2): 3—92.
Erdtman, G., Praglowski, J. and Nilsson, S., 1963. An Introduction to a Scandinavian Pollen Flora II, Almqvist and Wiksell, Uppsala, 89 pp.
Faegri, K. and Iversen, J., 1964. Textbook of Pollen Analysis. Munksgaard, Copenhagen, 2nd ed., 237 pp.
Fischer, C. A. H., 1890. Beiträge zur Morphologie der Pollenkörner. Dissertation, Breslau, 72 pp.
Keller, R., 1925. *Hypericum*. In: A. Engler and K. Prantl (Editors), Die natürlichen Pflanzenfamilien, 21, Engelmann, Leipzig, 2nd ed., pp. 175—183.
Khan, H. A., 1969. Pollen morphology of Indian Hypericaceae. J. Palynology, 5(2): 97—99.

Kuprianova, L. A. and Alyoshina, L. A., 1972. Pollen and Spores of Plants from the Flora of the European Part of the U.S.S.R. I. Izdatel'stvo "Nauka" Leningradskoe Otelenie, Leningrad, 171 pp. (in Russian).

Noack, K. L., 1939. Fortpflanzungsverhältnisse und Bastarde von *Hypericum perforatum* L. Z. Indukt. Abstammungs Vererbungsl., 76: 569—601.

Parmentier, P., 1901. Recherches morphologiques sur le pollen des Dialypétales. J. Botan., 15: 194—204.

Robson, N. K. B., 1957. Plant notes: *Hypericum maculatum* Crantz. Proc. Botan. Soc. Br. Isles, 2(3): 237—238.

Robson, N. K. B., 1968. *Hypericum*. In: T. G. Tutin, V. H. Heywood, N. A. Burges, D. M. Moore, D. H. Valentine, S. M. Walters and D. A. Webb (Editors), Flora Europaea, 2. Cambridge University Press, Cambridge, pp. 261—269.

Robson, N. K. B. and Adams, P., 1968. Chromosome numbers in *Hypericum* and related genera. Brittonia, 20(2): 95—106.

INDEX TO TAXA AND TERMS[1]

Adoxaceae, 6, 71
Adoxa moschatellina, 71, 72
Adoxa moschatellina type, 12, 71
Anagallis arvensis ssp. *arvensis*, 31, 36, 51
Anagallis arvensis ssp. *coerulea*, 31, 36, 51
Anagallis arvensis type, 34, 35, 43, 44
Anagallis crassifolia, 31, 36, 37, 51
*Anagallis foemina**, 31, 36
Anagallis minima, 31, 36, 37, 51
Anagallis tenella, 31, 36, 37, 51
Anagallis tenella type, 35, 36, 37, 44
Androsace carneo 32, 37, 38, 51
Androsace elongata, 32, 37, 38, 52
Androsace elongata group, 38
Androsace elongata type, 34, 37, 39
Androsace lactea, 32, 37, 38, 51
Androsace maxima, 32, 38, 52
Androsace maxima type, 33, 35, 38, 39
Androsace septentrionalis, 32, 37, 38
Androsace villosa, 32, 37, 38, 52
Apicalfeld, 31
Apocolpial field, 31
Apocolpium, 31
Asterolinon linum-stellatum, 32, 36, 37, 51

Blackstonia, 89
Blackstonia perfoliata ssp. *perfoliata*, 90, 92, 104
Blackstonia perfoliata ssp. *serotina*, 90, 92
Blackstonia perfoliata type, 92, 94, 96
Brassenia, 2
Bridge, 5, 31

Caprifoliaceae, 5
Caryophyllaceae, 1
Centaurium, 89
Centaurium erythraea, 90, 94, 104
Centaurium littorale, 90, 94, 95
Centaurium maritimum, 90, 94, 95
Centaurium pulchellum, 90, 94, 104
Centaurium pulchellum type, 92, 93, 94
Centaurium scilloides, 90, 94
Centaurium spicatum, 90, 93, 94
Centaurium tenuiflorum, 90, 94, 95
*Centunculus minimus**, 32, 36
Cicendia, 89

[1] Asterisks designate synonyms; roman numerals refer to pages of description.

Cicendia filiformis, 91, 95, 104
Cicendia filiformis type, 92, 93, 95, 96
Cortusa, 45
Cyclamen hederifolium, 32, 39, 52
Cyclamen hederifolium type, 34, 39, 40
Cyclamen purpurascens, 32, 40, 52
Cyclamen purpurascens type, 33, 39

Endocingulus, 31
Endocracks, 31

Fastigium, 5

Galinsoga parviflora, 2
Gentiana, 89
Gentiana Section *Aptera*, 97
Gentiana Section *Coelanthe*, 97
Gentiana Section *Cyclostigma*, 99
Gentiana Section *Pneumonanthe*, 97
Gentiana acaulis, 92, 98, 99
Gentiana asclepiadacea, 97
Gentianaceae, 89
Gentiana clusii, 92, 98, 99
Gentiana cruciata, 91, 96, 97, 105
Gentiana lutea, 91, 96, 97, 105
Gentiana nivalis, 91, 98, 99, 106
Gentiana pneumonanthe, 91, 96, 97, 104
Gentiana pneumonanthe type, 92, 96
Gentiana pannonica, 97
Gentiana purpurea, 91, 96, 97, 105
Gentiana verna, 91, 98, 106
Gentianella, 89, 102
Gentianella Section *Amarella*, 99
Gentianella Section *Crossopetalum*, 100
Gentianella amarella ssp. *amarella*, 91, 99, 100, 106
Gentianella amarella ssp. *septentrionalis*, 91, 99, 100
Gentianella amarella type, 99
Gentianella anglica, 91, 98, 99
Gentianella aurea, 91, 99, 100, 101, 102, 107
Gentianella campestris group, 98, 99
Gentianella campestris ssp. *baltica*, 91, 98, 99, 105
Gentianella campestris ssp. *campestris*, 91, 98, 99, 105
Gentianella campestris type, 92, 97, 99
Gentianella ciliata, 91, 98, 100, 105

Gentianella detonsa, *91, 100, 106*
Gentianella detonsa type, *91*, 100
Gentianella germanica, *91, 98, 99, 105, 106*
Gentianella tenella, *91, 99, 100, 101, 102, 107*
Gentianella type, *92*, 100
Gentianella uliginosa, *91, 98*
Glaux maritima, *32, 40, 52, 55*
Glaux maritima type, *35*, 40
Guttiferae, *125*

H-endoaperture, *89*
Horn, *31*
Hottonia palustris, *32, 41, 52, 55*
Hottonia palustris type, *33*, 41
Hypericum, *125, 132, 133*
Hypericum Section Brathys, *129*
Hypericum Section Hypericum, *133*
Hypericum androsaemum, *125, 127, 128*
Hypericum calycinum, *125, 127, 128, 140*
Hypericum calycinum type, *126*, 127, *128*
Hypericum canadense, *125, 129, 130, 140*
Hypericum canadense type, *126*, 128, *133*
Hypericum elegans, *125, 131*
Hypericum elodes, *125, 130, 132, 133, 140*
Hypericum elodes type, *126*, 130
Hypericum gentianoides, *126, 130*
Hypericum hircinum, *126, 127, 128, 140*
Hypericum hirsutum, *126, 131, 133*
Hypericum humifusum, *126, 131*
Hypericum humifusum group, *132, 133*
Hypericum inodorum, *126, 127, 128, 140*
Hypericum linarifolium, *126, 131*
Hypericum maculatum ssp. maculatum, *126, 132, 133, 141*
Hypericum maculatum ssp. obtusiusculum, *126, 132, 133*
Hypericum majus, *126, 129, 130*
Hypericum montanum, *126, 131, 132, 141*
Hypericum mutilum, *126, 129, 130, 140*
Hypericum mutilum group, *129*
Hypericum perforatum, *126, 132, 133, 141*
Hypericum perforatum group, *132, 135*
Hypericum perforatum type, *126*, 131, *132*
Hypericum pulchrum, *126, 132, 141*
Hypericum tetrapterum, *126, 132, 141*
Hypericum undulatum, *126, 132*

Linnaea borealis, *5, 7, 16*
Linnaea borealis type, *6*
Lomatogonium, *89*
Lomatogonium carinthiacum, *91, 103*
Lomatogonium rotatum, *91, 103, 107*
Lomatogonium rotatum type, *92, 102, 103*
Lonicera alpigena, *5, 7, 15*

Lonicera alpigena type, *6, 7*
Lonicera caprifolium, *5, 8, 15*
Lonicera caprifolium type, *6, 7, 8, 9*
Lonicera coerulea, *5, 9, 15*
Lonicera coerulea type, *6, 7, 8, 10*
Lonicera nigra, *5, 10, 15*
Lonicera periclymenum, *5, 9, 16*
Lonicera periclymenum type, *6, 7, 8, 9*
Lonicera xylosteum, *5, 10, 16*
Lonicera xylosteum type, *6, 7, 9, 10*
Lysimachia ephemerum, *32, 42, 53*
Lysimachia ephemerum type, *34, 35*, 41, 44
Lysimachia nemorum, *32, 37, 42, 54*
Lysimachia nemorum type, *34, 35, 36, 37, 42, 43*
Lysimachia nummularia, *32, 33, 43, 53, 55*
Lysimachia punctata, *32, 43, 44, 53*
Lysimachia terrestris, *32, 43, 44, 53*
Lysimachia thyrsiflora, *32, 43, 44, 53*
Lysimachia thyrsiflora group, *44*
Lysimachia vulgaris, *32, 43, 44, 53*
Lysimachia vulgaris group, *44*
Lysimachia vulgaris type, *33, 34, 42, 43, 44*

Microreticulate, *31*
Myrtus communis, *2*

Polar Area, *31*
Polarfeld, *31*
Primulaceae, *31*
Primula elatior, *32, 47, 48, 54, 55*
Primula farinosa, *32, 45, 52*
Primula farinosa type, *33, 44, 45, 46, 47*
Primula hirsuta, *32, 45, 54*
Primula hirsuta type, *34*, 45
Primula scotica, *32, 46, 54*
Primula scotica type, *33, 45, 46, 47*
Primula stricta, *32, 47, 52*
Primula stricta type, *33, 45*, 46
Primula veris, *32, 47, 48, 54*
Primula veris type, *33, 47*, 48
Primula vulgaris, *32, 47, 53, 54*
Prunus serotina, *2*

Retipilate, *5, 90*

Sambucus ebulus, *5, 11, 14, 15, 16*
Sambucus ebulus type, *6, 11, 13, 14*
Sambucus nigra, *5, 12, 15*
Sambucus nigra type, *6*, 11, *71*
Sambucus racemosa, *5, 12, 15*
Samolus valerandi, *32, 48, 55*
Samolus valerandi type, *34*, 48
Soldanella alpina, *32, 49, 55*

Soldanella alpina type, *33, 49*
Sparganiaceae, *75*
Sparganium angustifolium, 75, 77, 80, 81
Sparganium emersum, 75, 77, 80
Sparganium emersum type, *76, 78*
Sparganium erectum, 75, 78, 80
Sparganium erectum type, *76, 78, 79*
Sparganium erectum ssp. *erectum, 75, 79, 80*
Sparganium erectum ssp. *microcarpum, 75, 78, 79*
Sparganium erectum ssp. *neglectum, 75, 78, 79, 80*
Sparganium fluitans, 75, 77
Sparganium friesii, 75, 77, 81
Sparganium friesii group, *77, 78*
Sparganium glomeratum, 75, 77, 78, 81
Sparganium hyperboreum, 75, 77
Sparganium minimum, 75, 77, 81
Sparganium minimum group, *77, 78*
Steironema ciliata, 32, 49
Steironema ciliata type, *34, 43, 49*
Steironema lanceolata, 32, 49, 54
Striate, *90*

Striato-reticulate, *90*
Swertia, 89
Swertia perennis, 91, 101, 102, 107

Trientalis europaea, 32, 50, 54
Trientalis europaea type, *33, 34, 50*
Tsuga, 2
Typha angustifolia, 75, 76, 77, 80, 81
Typhaceae, *75*
Typha latifolia, 76, 80, 81
Typha latifolia type, *76, 79*
Typha laxmannii, 80
Typha minima, 76, 79, 80, 81
Typha shuttleworthii, 76, 80

Umbelliferae, *1*

Viburnum lantana, 5, 13, 14
Viburnum lantana type, *6, 11, 12, 14*
Viburnum opulus, 6, 13, 14, 16
Viburnum opulus type, *6, 11, 12, 13, 14*
Viburnum tinus, 6, 14
Viburnum tinus type, *6, 11, 12, 13*

Errata

p. *31.* Under specimens examined:*Anagallis arvensis* L. ssp. *coerulea* (Gouan) Vollmer — read *Anagallis foemina* (Miller) Schinz et Thellung.

p. *78.* Line 4: *Minimum* group — read *S. minimum* group.
Line 7: *Friesii* group — read *S. friesii* group.

p. *79.* Line 14: *S. emersum* type (p. *78*) — read *S. emersum* type (p. *76*).

p. *104.* Line 14: endexine — read nexine.